Environmental Science and Engineering

Series Editors

The ultimate goal of this series is to contribute to the protection of our environment, which calls for both profound research and the ongoing development of solutions and measurements by experts in the field. Accordingly, the series promotes not only a deeper understanding of environmental processes and the evaluation of management strategies, but also design and technology aimed at improving environmental quality. Books focusing on the former are published in the subseries Environmental Science, those focusing on the latter in the subseries Environmental Engineering.

Nezha Mejjad
Editor

Nuclear-Based Approaches application for marine pollution assessment and monitoring

Springer

Editor
Nezha Mejjad
National Centre for Nuclear Energy
Science and Technology (CNESTEN)
Rabat, Morocco

ISSN 1863-5520 ISSN 1863-5539 (electronic)
Environmental Science and Engineering
ISBN 978-3-032-12196-7 ISBN 978-3-032-12197-4 (eBook)
https://doi.org/10.1007/978-3-032-12197-4

This Springer imprint is published by the registered company Springer Nature Switzerland AG
The registered company address is: Gewerbestrasse 11, 6330 Cham, Switzerland

Contents

Chapter 1
Nuclear-Based Approaches for Marine Pollution Assessment: A Bibliometric Analysis

Nezha Mejjad and Hicham Boutracheh

Abstract Nuclear-based approaches offer powerful tools for monitoring and managing ocean pollution through accurate tracing and assessment capabilities. The present study focuses on analyzing publications trends and studies that used isotopes and nuclear technologies for marine pollution assessment and management. In addition, the use of nuclear approaches for microplastic monitoring, chemical contaminants assessment, blue carbon sequestration rates estimation, oil spill tracking, and ocean acidification tracking over time was reviewed and analyzed. The chapter showed the critical role that nuclear technology plays in advancing knowledge and enhancing understanding of marine pollution. The bibliometric analysis underscores the lack of studies related to nuclear approaches as a potential solution to tackle the current marine pollution issues while the focus is mainly on atmospheric fallout of radioelement and the use of nuclear techniques for chemical pollution assessment using sediment cores, radiation levels, absorbed dose and limit values. The study analysis calls to improve investment in nuclear-based research, which can contribute to sustainable solutions for preserving and sustaining ocean health for future generations.

Keywords Nuclear-based approaches · Marine pollution · Microplastic · Oil spills · Heavy metals · Blue carbon · Bibliometric analysis

N. Mejjad (✉)
National Centre for Nuclear Energy, Science and Technology (CNESTEN), Rabat, Morocco

H. Boutracheh
Laboratory of Engineering Sciences, National School of Applied Sciences, Ibn Tofail University, Kénitra, Morocco

CNRST, IMIST, National Center for Scientific and Technical Research, Rabat, Morocco

N. Mejjad (ed.), *Nuclear-Based Approaches application for marine pollution assessment and monitoring*, Environmental Science and Engineering,
https://doi.org/10.1007/978-3-032-12197-4_1

1.1 Introduction

Marine pollution is among the most discussed and investigated topics. Various types of pollution influence Coastal and marine areas, threatening life below water. The development of human activities increased ocean vulnerability mainly due to overfishing (Sumaila and Tai 2020; de Mitcheson et al. 2020; Ling et al. 2009), waste dumping (Graham 2024; Hazarika 2023; Hamblin 2008), and natural resources exploitation (oil and gas; (Mein et al. 2021; Andrews et al. 2021; Lawson et al. 2022)). This has serious economic, ecological, and health impacts (Fig. 1.1).

Since the 1970s, there has been an increasing tendency towards conserving the ocean and life below water through increasing awareness among different components of society to avoid the loss of marine life in coastal areas (e.g., bays and

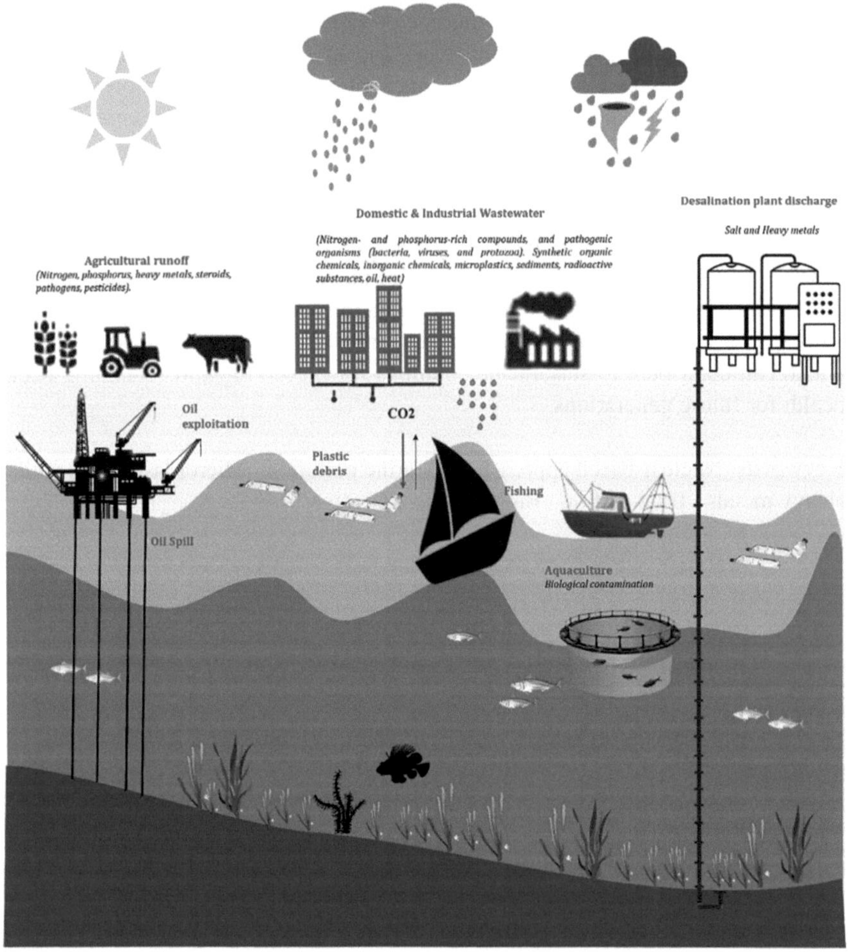

Fig. 1.1 Different human activities influencing oceans

estuaries). Until the 1970s, the ocean was a legal dumping site for different types of waste (industrial, industrial, nuclear, and other waste, nuclear, and other types of waste) (the United States Environmental Protection Agency (EPA)). Then, waste dumping in the ocean was banned in response to organized congresses that brought coastal and marine ecosystem protection and conservation to the forefront of environmental debates. In 1972, the "London Convention" was signed to protect marine systems from waste dumping in the ocean. Also, since this date, there has been an increasing trend toward assessing and monitoring pollution levels and marine environment quality worldwide (Mejjad et al. 2023).

Nonetheless, the deposition of waste into the ocean continues to occur in various forms globally (Hammer et al. 2012; Clapp 2002). Research conducted in recent decades indicates that the degradation of oceanic environmental quality is predominantly attributable to the influences of human activities and the effects of climate change (Wang et al. 2024; Wei et al. 2024; Wang and Gu 2021; Gattuso et al. 2018; Cloern et al. 2016; Hoegh-Guldberg and Bruno 2010; Whitehead et al. 2009).

Accordingly, scientific research advancement allowed for a better understanding of marine pollution, and thus, the body of research has provided critical information and data that informs public awareness and policy-making. Various approaches were used for this purpose, including nuclear approaches, which provided valuable tools to address ocean pollution through different methodologies and techniques that contribute to assessing, monitoring, understanding, and mitigating the impacts of pollutants (Ghanem et al. 2024; Gore et al. 2024; Wu et al. 2024; Yang et al. 2024; Benmhammed et al. 2021; Uddin et al. 2021; Sefton et al. 2020; Gailis et al. 2021; Taieb Errahmani et al. 2020; Mejjad et al. 2018; Maher et al. 2017);

The present chapter is the first assessment of research records and trends related to applying nuclear-based approaches to marine pollution analysis and monitoring. In this sense, a bibliometric analysis is carried out to establish map visualization networks of terms related to the search query "Nuclear-based approaches application for marine pollution analysis and monitoring". The study thus sought to show the current advancement in employing these tools for researching ocean environmental issues monitoring, such as heavy metal pollution, microplastic pollution, and blue carbon sequestration.

1.2 An Overview of Nuclear Approaches in Service of Marine Environment Protection

For decades, nuclear approaches have been utilized to understand and identify various processes that govern the oceans, including the monitoring and control of ocean ecosystem health. Applying nuclear methods and their derived approaches aids in decision-making and developing management strategies, proposing essential measures for safeguarding and protecting the marine environment from different forms of pollution, climate change impacts, and habitat and biodiversity loss.

Around 80% of pollutants reach oceans via land-related activities (Daoji and Daler 2004). Various hazardous substances enter coastal and marine environments through surface water and groundwater, such as persistent organic compounds (POC), including industrial chemicals and pesticides, hydrocarbons, heavy metals, and radioactive substances. To address pollution in coastal and marine environments, various science-based tools and techniques have been developed by the International Atomic Energy Agency (IAEA) to monitor, protect, and minimize the impacts of pollutants on the marine environment (IAEA 2021a).

1.2.1 Nuclear Approaches for Pollutants Monitoring

Nuclear methodologies are used to trace the origin and history of pollutants, which would help assess their impact and establish remediation and mitigation strategies. Natural and artificial radionuclides are employed as tracers to establish the chronology of pollution events and thus identify the pathways of contaminants into the ocean, which is important. As stated before, 80% of pollutants in the marine environment enter from land (Daoji and Daler 2004), including domestic and industrial discharges and agricultural runoff.

Marine sediment cores are used to assess the vertical distribution of pollutants such as radioactive elements (Ghanem et al. 2024; Benmhammed et al. 2021; Laissaoui et al. 2018; Klubi et al. 2017; Mejjad et al. 2016), heavy metals (Mejjad et al. 2018), microplastics (Lanctôt et al. 2018) and also for estimating the carbon accumulation rates (Choudhary et al. 2024; Fu et al. 2023). Using radiotracers for sediment core analysis, it is possible to trace pollutants' deposition history, sediment accumulation rate, and sedimentation rates over time. This allows for understanding the changes in sediments' hydrodynamic and deposition conditions, tracking pollution history, and thus defining possible sources of pollutants. Ghanem et al. (2024) assessed the history of heavy metal input using radio-dating of lagoonal coastal sediment cores, which revealed a significant increase in heavy metal content following the development of human activities, including growth in urbanization and industrial activities. Thus, the application of nuclear approaches enables the establishment of a timeline for pollutant deposition and allows for the identification of heightened concentration periods for heavy metals. Mejjad et al. (2018) reported the same finding, where the increase in metal concentrations was correlated well with the development of human activities in the studied coastal lagoon.

1.2.2 Isotopes and Nuclear Approaches for Microplastics Pollution, Bioaccumulation, and Biodistribution Tracking

Isotopes and nuclear techniques are also used to assess microplastic (MP) distribution trends over time and evaluate their potential impacts. Radiometric dating can also be used to trace the possible origins of plastics by assessing sediment cores (Torres and De-la-Torre 2021; Bancone et al. 2020; Turner et al. 2019). MPs threaten marine organisms because of their persistence and capacity to accumulate toxic substances. Vidal et al. (2024) have used sediment cores to examine the contamination trends dating back to 1910 for the Loire River by the mean of ^{210}Pbex and ^{137}Cs isotopes, which allowed for an understanding of the historical impact of plastic additives on river ecosystems. Besides, radiotracers can be used to analyze how MP interacts with pollutants and their possible transfer through the food chain. Munir et al. (2023) have developed a technique for preparing Iodine-131 (I-131) labeled PVC to investigate its feasibility as a tool for tracking microplastic movement and accumulation in biological tissues. Nuclear tracing approaches have advantages over traditional methods, and their incorporation allows for yielding crucial data and information on how MPs interact with biological systems by understanding their pathways in food webs and, thus, their potential effects on human health.

Besides, it was reported that MPs in coastal and marine environments could absorb and carry on harmful chemicals such as PCBs (polychlorinated biphenyls) (Smith et al. 2018; Rafa et al. 2024), which marine organisms may subsequently ingest (Smith et al. 2018; Devriese et al. 2017). Using radiotracers, it is possible to examine the pathways and bioaccumulation of pollutants in marine ecosystems (Munir et al. 2023; Pyl et al. 2022). Pyl et al. (2022) presented the first experimental demonstration of how MPs can mediate the transfer of PCB-153 into sea urchin tissues using radiolabeled ^{14}C-PCB-153 adsorbed onto MPs. The obtained results revealed the role the MP in seafloor plays as vectors for PCB-153 transfer to adult sea urchin tissues. Accordingly, the use of radiotracer tools allows for a better understanding of harmful contaminants' movements and pathways to marine organisms.

1.2.3 Isotopes and Nuclear Approaches for Microplastic Recycling

^{13}C is a stable carbon isotope, a valuable tool for differentiating between various sources of plastics and evaluating their degradation (Berto et al. 2019; Taipale et al. 2019). Based on recent studies, the variation of the isotopic composition of plastics can allow for defining occurring processes of biotic and abiotic degradation (Birch et al. 2020; Kuznetsova et al. 2023) in marine environments, which helps establish the recycling strategies for plastics. The implementation of radiation technologies can enhance plastic recycling. In 2020, IAEA initiated an initiative called NUTEC Plastics (NUclear TEchnology for Controlling Plastic Pollution) to tackle plastic pollution, including two fields: control and impact assessment of MPs in the coastal and marine environments and enhancing recycling processes of plastic waste using radiation technologies (Ponomarev et al. 2022; IAEA 2021b). Thus, nuclear approaches complement traditionally used methods for plastic recycling, consequently enhancing conventional recycling practices and renovating end-of-life plastic by transforming plastic-related waste into reusable resources (IAEA 2021b). Radiation technology based on the IAEA report (IAEA 2021a) is filling existing gaps in handling plastic waste; for instance, this technology will solve challenges related to sorting plastic waste, which enhances the value and quality of recycled plastics. In addition, the radiation can be utilized to decompose or transform waste plastic polymers into fuel or smaller components, which can then serve as chemical feedstocks for the production of consumer products, either with or without the incorporation of virgin materials (Ponomarev et al. 2022; Trajano 2021). Accordingly, nuclear technology can tackle microplastic pollution, promote a circular economy, and allow greener and cleaner production.

1.2.4 Nuclear Approaches for Oil Spill Management in Ocean

Nuclear approaches have emerged as valuable tools for managing ocean oil spills. This technology provides the ability to trace oil spill sources, evaluate environmental impacts, and thus support mitigation efforts (Bastos et al. 2022; Patterson III et al. 2020; Barrosa et al. 2013; Betti et al. 2011). Patterson III et al. (2020) have analyzed the use of stable and radioisotopes to trace the impacts of marine oil spills on food webs. The study demonstrates the usefulness of isotopes, mainly ^{13}C and ^{15}N, for deducing changes in food web resource availability and structure following oil spills as they can reveal shifts in the position of trophic and petrochemicals assimilation within coastal and marine ecosystems. Also, the study emphasized the importance of these isotopes as long-term biogeochemical tracers that offer information about hydrocarbon assimilation and transfer through marine food webs. The depletion of hydrocarbon value in fish tissues indicates petrocarbon incorporation into the food web. This is crucial for understanding the ecological footprint of oil spills (Patterson III et al. 2020). Nuclear techniques can provide valuable tools in oil spill events for assessing the extent of contamination and identifying spill sources using gas-chromatography mass-spectrometry. This latter analyzes oil samples and allows for developing chemical "fingerprints" for tracing oil back to its source (Rabinovitch et al. 2024; Zhang et al. 2021; Yang et al. 2014; Wang et al. 2006). Nuclear isotopes also allow for biodegradation processes monitoring by tracking the natural occurrence and abundance of elements such as hydrogen, carbon, and oxygen implicated in hydrocarbon breakdown (Musat et al. 2016; Vieth and Wilkes 2010; Prinzhofer and Battani 2003; Conrad et al. 1999). This approach is essential to quantify the biodegradation rates and pathways as it allows for the disclosure of remediation approaches. By understanding the biodegradation processes, scientists can improve cleanup efforts and decrease costs related to engineered remediation.

1.2.5 Nuclear Approaches for Addressing Ocean Acidification

The purpose of using nuclear technology to address ocean acidification is to provide a source of low-carbon energy that allows for minimizing greenhouse gas emissions. Nuclear methods have been widely used to study changes in ocean acidity over the years and their potential impacts on marine organisms (IAEA 2019). Nuclear power has played a crucial role in the past 50 years by contributing to CO_2 emissions reduction by 60 Gt (IAEA 2020). Besides, using boron isotopes allows it to monitor and assess how the ocean changes, which allows for assessing historical ocean pH levels and thus determining past acidification events (Donald 2017; Xiao et al. 2013). This technique consists of using the boron isotopic composition originating in marine carbonates, such as in the coral reefs and fossilized organisms, to establish seawater acidity history (Piazza 2023; Hoefs and Harmon 2023; Levin et al. 2015). Accordingly, nuclear technology contributes indirectly to decreasing ocean acidification.

1.2.6 Blue Carbon

Blue Carbon is the captured carbon stored by coastal ecosystems known as Coastal Blue Carbon Ecosystems (CBCE). These ecosystems include seagrass, mangroves, and salt marshes, which are critical in climate change mitigation. CBCEs sequester Carbon dioxide and store it in their biomass and sediments (Williamson and Gattuso 2022; Hori et al. 2019; Howard et al. 2018). Nuclear technology allows a better understanding of carbon accumulation rates over time. Using isotopes such as Pb-210 and Cs-137, it is possible to determine the amount of sequestered carbon in specific zones over decades and centuries (Khodadadi et al. 2024; Dinsley et al. 2018; Arias-Ortiz et al. 2018) and track sediment composition change and human activities development over time. Integrating nuclear techniques into blue carbon subjects can provide robust data supporting the management of blue ecosystems (Rogers et al. 2023; Lamont 2022; Kelleway et al. 2017).

1.3 Bibliometric Analysis of Research Trends

1.3.1 Data Collection

Data were collected from the Dimensions.ai database (accessed on January 10, 2025), which includes papers, chapters, edited books, preprints, and proceedings. The search parameters query are as follows:

- *Data range*: 1933–2025
- *Document type*: Articles, proceedings, chapters, edited books, and preprints (Fig. 1.2).
- *Search query date*: 10th January 2025.
- *Used terms*: Pollution AND Marine AND/OR Coastal AND Radioactive OR Radionuclide.
- *The number of extracted publications*: 2500 publications.

1.3.2 Research Category

Twenty-two thousand four hundred and nineteen documents related to our search query were published between 1933 and 2025 based on data found in Dimensions.ai. Most of these publications were related to the field category (Based on ANZSRC 2020 indicator, Fig. 1.3) "Environmental Sciences, with 4504 publications, followed by Earth Science (3606 publications), Biological Sciences (1439 documents),

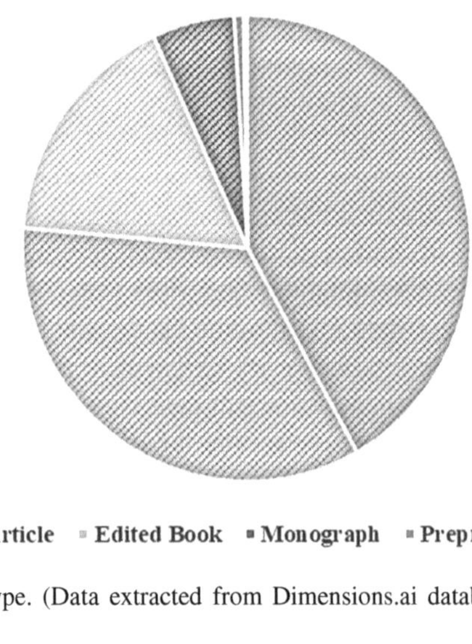

Fig. 1.2 Publication type. (Data extracted from Dimensions.ai database (accessed on January 10, 2025))

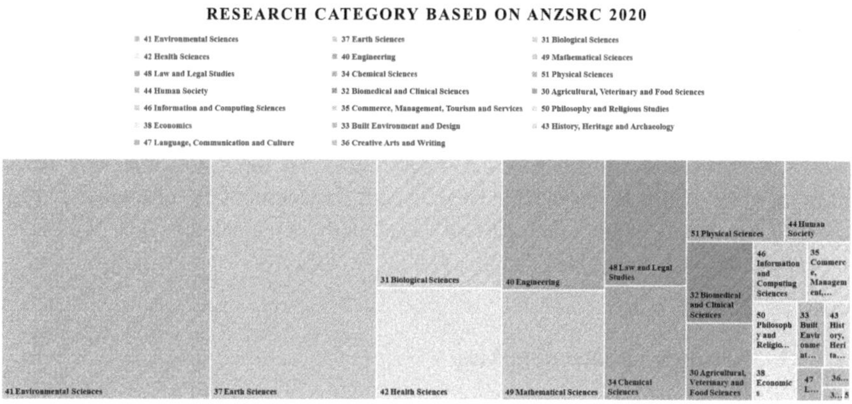

Fig. 1.3 Research category based on ANZSRC 2020

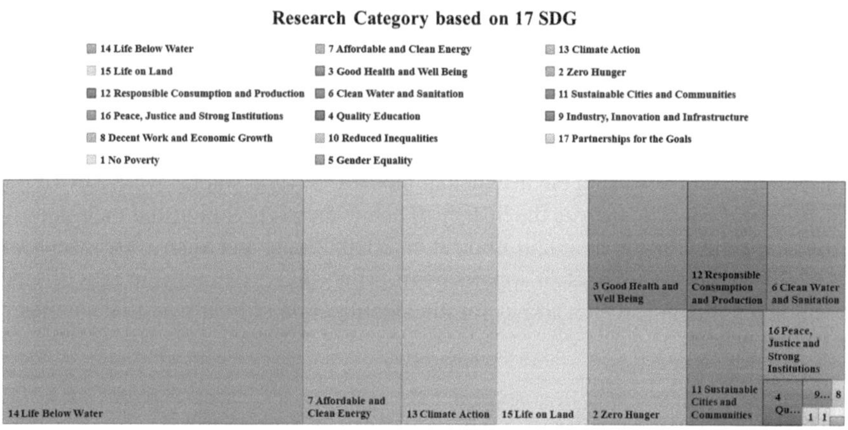

Fig. 1.4 Research category based on 17SDG

and Health Sciences (1244 documents)" (Fig. 1.3). This distribution indicates that the literature related to our search query mainly focuses on environmental, earth sciences, biological science, and health. It should be noted that natural and artificial radionuclides are significantly applied in coastal and marine environments and serve as valuable tools for tracking pollution (Benmhammed et al. 2021; Uddin et al. 2021; Mejjad et al. 2018, 2022) understanding oceanographic processes (Qiao et al. 2023) and monitoring and assessing ecological impacts (Lind et al. 2013).

Based on the 17 Sustainable Development Goals (17SDGs), the search query-related literature revealed that the main SDG research category related to the use of nuclear approaches for marine and coastal pollution assessment is life below water (14SDG), Affordable and clean energy (SDG7) and Climate Action (13SDG) (Fig. 1.4).

1.3.3 Bibliometric Analysis

Two thousand five hundred documents were extracted from the dimensions.ai database. The documents were sorted by relevance to our search query. Of 48,904 keywords, 735 meet the threshold of 20 occurrences. At the same time, 441 is the number of keywords to be selected by relevance, presenting 60% of total keywords meeting the threshold and most relevant based on the score calculated by Vosviewer (Van Eck and Waltman 2022).

1.3.3.1 Network Visualization Map

The Network visualization map shows five clusters (Fig. 1.5). The red and green clusters indicate 95 items or keywords for each. Eighty keywords are presented in blue color while the yellow and purple clusters are shown each 79 items. The colored nodes and circles are linked by 48,013 links and a total link strength of 150,841.

The most occurred keywords in the ***red cluster*** include heavy metal (co-occurrence: 257 items), metal (co-occurrence: 312), index (co-occurrence: 173), sediment sample (co-occurrence: 148 items), enrichment (co-occurrence: 111). This cluster is mainly related to analyzing and assessing heavy metal concentrations in sediment samples using environmental indexes. Several studies were carried out on sediment cores to analyze the history of heavy metals pollution and their deposition over time (Tang et al. 2022; Zhou et al. 2020; Zhang and Shan 2008; Meybeck et al. 2007). The use of nuclear approaches is primordial in defining the ages of sediment and pollutants, which allows for the identification of their potential sources.

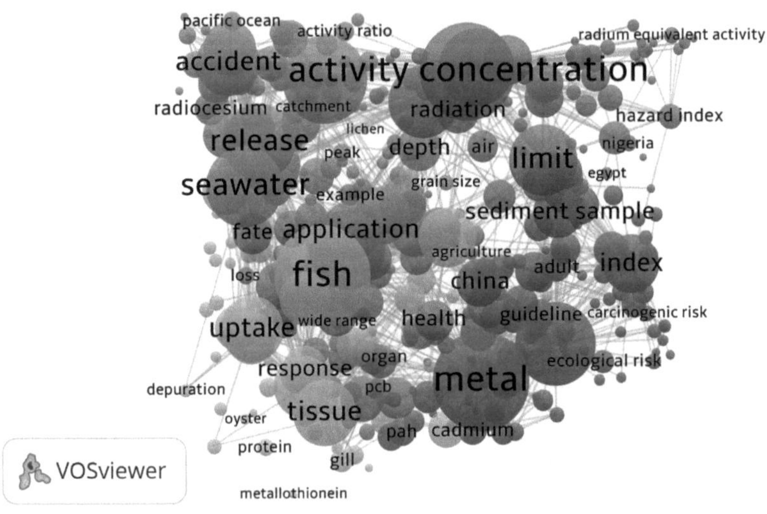

Fig. 1.5 Map network visualization of main keywords related to the search query

The ***green cluster*** covers keywords related to radioactive materials levels measurement, such as ^{137}Cs (co-occurrence: 273) in seawater (co-occurrence: 248), which serves as the medium for pollutant transport (co-occurrence: 184) through marine environments. At the same time, nuclear power plants (co-occurrence: 164) and nuclear accidents (co-occurrence: 148) denote unintentional releases (co-occurrence: 229) that can lead to significant environmental impacts. Fallout term refers to radioactive particle deposition from the atmosphere.

The ***blue cluster*** includes keywords such as activity concentration (co-occurrence: 307), dose (co-occurrence: 217), dose rate (co-occurrence: 146), limit (co-occurrence: 203), radiation (co-occurrence: 127), natural radionuclide (co-occurrence: 111), ingestion (co-occurrence: 94). These keywords underline the relationship between nuclear technology and marine environmental safety. The activity concentration is the measure of the radioactive content in a specific volume, while the dose signifies radiation absorbed by marine organisms. Thus, this cluster covers topics related to radiation exposure and regulatory thresholds for permissible radiation doses. Collectively, these terms underscore the need for monitoring radiological risks and establishing practical management approaches to protect and conserve marine ecosystems and public health from potential radioactive contamination.

The ***yellow cluster*** covers keywords related to radioactive contamination level analysis and assessment in fish species. The main items or keywords includes fish (co-occurrence: 314), consumption (co-occurrence: 175), uptake (co-occurrence: 197), tissue (co-occurrence: 197), bioaccumulation (co-occurrence: 185), food (co-occurrence: 158), transfer (co-occurrence: 139), muscle (co-occurrence: 111), fish species (co-occurrence: 87). The keywords "fish" mainly was occurred suggesting the interest in aquatic life conservation and monitoring through analyzing uptake and consumption rates, which refer to assessing how pollutants including radionuclides are absorbed and ingested by marine organisms. This is possible by analyzing radionuclides activity concentrations in fish tissues mainly the liver and muscles (e.g., (Barescut et al. 2011; Sirin 2019)), where they accumulate through a process called bioaccumulation. These term occurrences in common clusters show pathways through which radioactive contamination can disturb life below water and affect human health. This highlights the need for effective and continuous monitoring and management strategies in marine pollution perspectives.

The ***purple cluster*** encompasses keywords related to nuclear approaches application (co-occurrence: 176), mechanism (co-occurrence: 136), needs (co-occurrence: 123), understanding (co-occurrence: 97), knowledge (co-occurrence: 103), toxicity (co-occurrence: 117), and health (co-occurrence: 114). These different keywords are related to studies that apply nuclear methods to address marine pollution issues. At the same time, they show the need for increasing knowledge and understanding in the field (EI-Motaium 2007). The occurrence of terms such as health and toxicity suggests the increasing attention toward monitoring the health potential risks associated with pollution and the urgent need for more comprehensive assessment and evaluation of pollution levels to protect and conserve marine life and human health. These keywords reflect the need to effectively integrate nuclear technology into marine pollution management programs and strategies.

1.3.3.2 Overlay Visualization Map

Figure 1.6 displays the overlay visualization map designed to generate additional insight into the existing relationships between keywords, authors, and journals. This visualization allowed us to determine publication trends over time. The average publication year of keywords is represented by colors, where yellow indicates keywords related to more recent published studies, and blue signifies terms associated with older publications.

The map indicates that current research trends focus on analyzing radionuclides' effects on human health by using indexes such as the hazard index, carcinogenic index, and radium equivalent activity. Between 2012 and 2014, research was mainly focused on analyzing concentration levels of pollutants, including metals and radionuclides, in fish tissues. Also, the circle size denotes the importance of keywords within the dataset. "Fish" and "Seawater" were the most occurred keywords in publications published between 2012 and 2014. In 2020, the keywords with high occurrence frequency were "index" and "hazard index" reflecting the current trends toward analyzing the chemical pollutant effects on human health.

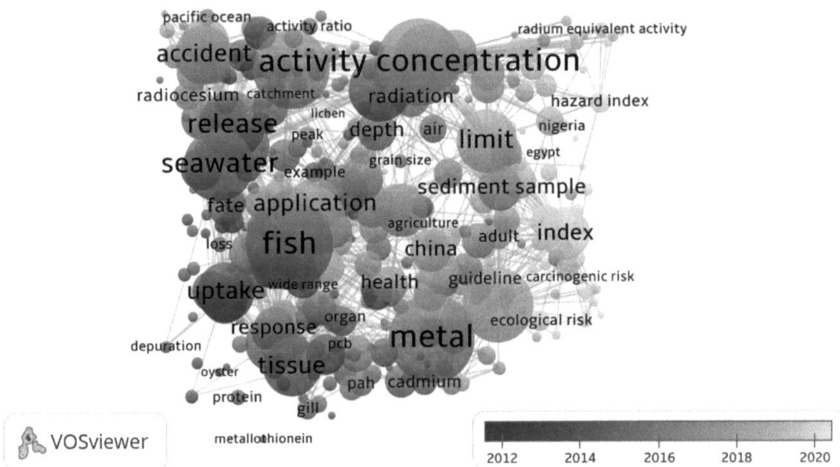

Fig. 1.6 Map overlay visualization of main keywords related to the search query

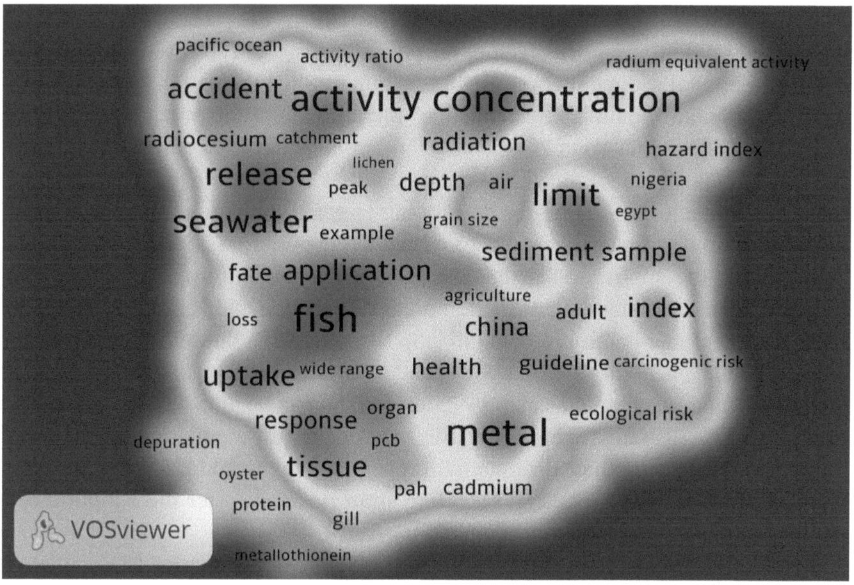

Fig. 1.7 Heat map or density map of main keywords related to the search query

1.3.3.3 Density Visualization Map

The density or heat map further demonstrates the prominence of specific items (keywords) (Fig. 1.7). The density of keywords is represented by colors ranging from blue and green to yellow and red. Keywords shown in red and yellow areas signify important weight and a high number of neighboring items; the occurrence of few items in the neighborhood characterizes those close to blue and green colors. Thus, the visualization map underlines the significance of "fish", "activity concentration", "index", "sediment sample", "metal", "release", "uptake", "seawater", "tissue," and "limit" as hot topics within research related nuclear approaches for marine pollution management.

1.4 Analysis and Discussions

The chapter underlined the importance of nuclear-based methodologies in increasing our understanding and managing marine pollution. The bibliometric analysis indicated a growing body of literature that emphasizes different uses of nuclear approaches for pollutant monitoring, origins assessment, analysis, and evaluation of their potential impact on marine ecosystems. However, despite the advancements in this field, a notable gap in research related to nuclear approach technologies to address the pressing challenges of marine pollution was noticed.

This review analysis shows the effectiveness of nuclear approaches in identifying the origin of pollution by establishing the pollution history. The application of radio dating for coastal sediment cores enables estimating sedimentation rates and sediment accumulation rates and establishing the chronology of pollutant deposition and defining pollutant sources, thereby linking contaminant content in sediment with the development of various human activities over the years. This temporal trend determination is crucial for developing targeted mitigation plans and strategies and informing policy decisions for remediating pollution impacts.

Similarly, the application of nuclear methods for monitoring MPs allows for advancing our understanding of the temporal distribution and ecological effects of these emergent and persistent contaminants. Tracing MPs through food webs and assessing their interactions with toxic substances provides critical data for evaluating risks to marine organisms and, by extension, human health.

This chapter also highlights innovative methods for differentiating the source of plastics and assessing their degradation processes. This would allow the development of effective and applicable recycling strategies and foster the adoption of a circular economy in plastic management.

In addition, nuclear approaches' ability to address various marine issues, including oil spills and ocean acidification, reveals its versatility in tackling diverse aspects of marine pollution. These methods can provide important insight by tracing oil spill sources and monitoring biodegradation processes, which can enhance cleanup efforts and reduce remediation-related costs.

1.5 Conclusions

The present chapter analyzed the role of nuclear approaches in marine pollution assessment, monitoring, and management. However, the bibliometric analysis showed that the collected studies mainly focused on evaluating radioactive contamination levels. At the same time, nuclear approaches as tools for pollution monitoring and management remain understudied and/or not fully understood.

In addition, based on the visualization map, the use of nuclear approaches for tackling microplastic pollution and mitigating climate change effects in the framework of blue carbon needs to be addressed as most of the studies were related to atmospheric fallout of radioelement and use of nuclear approaches for chemical pollution assessment using sediment cores, radiation levels, absorbed dose and limit values.

This chapter's findings thus highlight the need for further studies emphasizing the potential roles of nuclear approaches in marine pollution mitigation and management. However, this chapter would contribute to the existing knowledge and highlight the key areas of nuclear approaches for marine pollution management.

By fostering interdisciplinary research and promoting collaboration among researchers, environmental managers, and policymakers to leverage nuclear technologies and approaches in marine conservation efforts, it is possible to improve our ability to save and protect aquatic ecosystems from different forms of marine pollution.

References

Andrews N, Bennett NJ, Le Billon P, Green SJ, Cisneros-Montemayor AM, Amongin S et al (2021) Oil, fisheries and coastal communities: a review of impacts on the environment, livelihoods, space and governance. Energy Res Soc Sci 75:102009
Arias-Ortiz A, Masqué P, Garcia-Orellana J, Serrano O, Mazarrasa I, Marbà N et al (2018) Reviews and syntheses: ^{210}Pb-derived sediment and carbon accumulation rates in vegetated coastal ecosystems—setting the record straight. Biogeosciences 15(22):6791–6818
Bancone CE, Turner SD, Ivar do Sul JA, Rose NL (2020) The paleoecology of microplastic contamination. Front Environ Sci 8:574008
Barescut J, Lariviere D, Stocki T, Wesley SG, Khan MF (2011) Radionuclides in south Indian seafoods: a special focus on major contributing species. Radioprotection 46(6):S151–S159
Barrosa VG, Oliveiraa TMN, Vaza C, Zuppi GM (2013) Lead and stable isotopes in sediments of Babitonga Bay: an oil spill case. J Environ Hydrol 21:1–10
Bastos LPH, da Costa Cavalcante D, Alferes CLF, da Silva DBN, de Oliveira Ferreira L, Rodrigues R, Pereira E (2022) Fingerprinting an oil spill event (August of 2021) in the oceanic Fernando de Noronha archipelago using biomarkers and stable carbon isotopes. Mar Pollut Bull 185:114316
Benmhammed A, Laissaoui A, Mejjad N, Ziad N, Chakir E, Benkdad A, Bouh HA, El Yahyaoui A (2021) Recent pollution records in Sidi Moussa coastal lagoon (western Morocco) inferred from sediment radiometric dating. J Environ Radioact 227:106464
Berto D, Rampazzo F, Gion C, Noventa S, Formalewicz M, Ronchi F et al (2019) Elemental analyzer/isotope ratio mass spectrometry (EA/IRMS) as a tool to characterize plastic polymers in a marine environment. In: Plastics in the environment, pp 37–54
Betti M, Boisson F, Eriksson M, Tolosa I, Vasileva E (2011) Isotope analysis for marine environmental studies. Int J Mass Spectrom 307(1–3):192–199
Birch QT, Potter PM, Pinto PX, Dionysiou DD, Al-Abed SR (2020) Isotope ratio mass spectrometry and spectroscopic techniques for microplastics characterization. Talanta 224:121743
Choudhary B, Dhar V, Pawase AS (2024) Blue carbon and the role of mangroves in carbon sequestration: its mechanisms, estimation, human impacts and conservation strategies for economic incentives. J Sea Res 199:102504
Clapp J (2002) The distancing of waste: overconsumption in a global economy. The MIT Press
Cloern JE, Abreu PC, Carstensen J, Chauvaud L, Elmgren R, Grall J et al (2016) Human activities and climate variability drive fast-paced change across the world's estuarine–coastal ecosystems. Glob Chang Biol 22(2):513–529
Conrad ME, Templeton AS, Daley PF, Alvarez-Cohen L (1999) Isotopic evidence for biological controls on migration of petroleum hydrocarbons. Org Geochem 30(8):843–859
Daoji L, Daler D (2004) Ocean pollution from land-based sources: East China Sea, China. Ambio:107–113
de Mitcheson YJS, Linardich C, Barreiros JP, Ralph GM, Aguilar-Perera A, Afonso P et al (2020) Valuable but vulnerable: over-fishing and under-management continue to threaten groupers so what now? Mar Policy 116:103909
Devriese LI, De Witte B, Vethaak AD, Hostens K, Leslie HA (2017) Bioaccumulation of PCBs from microplastics in Norway lobster (Nephrops norvegicus): an experimental study. Chemosphere 186:10–16
Dinsley JM, Gowing CJB, Marriott AL (2018) Natural and anthropogenic influences on atmospheric Pb-210 deposition and activity in sediments: a review. British Geological Survey, Nottingham, UK, p 27
Donald H (2017) Boron isotopes as a proxy for pH in siliceous and calcareous marine algae. Doctoral dissertation, University of Southampton
El-Motaium RA (2007) Application of nuclear techniques in environmental studies and pollution control. https://inis.iaea.org/records/g79ax-rj122/files/39038848.pdf

Fu C, Frappi S, Havlik MN, Howe W, Harris SD, Laiolo E, Gallagher AJ, Masqué P, Duarte CM (2023) Substantial blue carbon sequestration in the world's largest seagrass meadow. Commun Earth Environ 4(1):1–9

Gailis M, Kohfeld KE, Pellatt MG, Carlson D (2021) Quantifying blue carbon for the largest salt marsh in southern British Columbia: implications for regional coastal management. Coast Eng J 63(3):275–309

Gattuso JP, Magnan AK, Bopp L, Cheung WW, Duarte CM, Hinkel J et al (2018) Ocean solutions to address climate change and its effects on marine ecosystems. Front Mar Sci 5:337

Ghanem A, Nada A, Abu-Zeid H, Madcour W, Shetaia SA, Imam N (2024) Historical trends of heavy metals applying radio-dating and neutron activation analysis (NAA) in sediment cores, Burullus Lagoon, Egypt. Environ Sci Pollut Res 31(31):43633–43658

Gore C, Gehrels WR, Smeaton C, Andrews L, McMahon L, Hibbert F, Austin WE, Nolte S, Garrett E (2024) Saltmarsh blue carbon accumulation rates and their relationship with sea-level rise on a multi-decadal timescale in northern England. Estuar Coast Shelf Sci 299:108665

Graham RE (2024) A study of illegal dumping near coastal zones and waterways in Grenada: advancing waste management solutions and reducing marine litter. PLOS Water 3(4):e0000175

Hamblin JD (2008) Poison in the well: radioactive waste in the oceans at the dawn of the nuclear age. Rutgers University Press

Hammer J, Kraak MH, Parsons JR (2012) Plastics in the marine environment: the dark side of a modern gift. Rev Environ Contam Toxicol:1–44

Hazarika R (2023) Exploring the potential impacts of waste disposal sites on ocean ecosystem contamination in Newfoundland: a geospatial analysis and public perception study. Doctoral dissertation, Memorial University of Newfoundland

Hoefs J, Harmon RS (2023) Isotopic history of seawater: the stable isotope character of the global ocean at present and in the geological past. Isot Environ Health Stud 59(4–6):349–411

Hoegh-Guldberg O, Bruno JF (2010) The impact of climate change on the world's marine ecosystems. Science 328(5985):1523–1528

Hori M, Bayne CJ, Kuwae T (2019) Blue carbon: characteristics of the ocean's sequestration and storage ability of carbon dioxide. In: Blue carbon in shallow coastal ecosystems: carbon dynamics, policy, and implementation. Springer, pp 1–31

Howard JL, Creed JC, Aguiar MV, Fourqurean JW (2018) CO_2 released by carbonate sediment production in some coastal areas may offset the benefits of seagrass "Blue Carbon" storage. Limnol Oceanogr 63(1):160–172

IAEA (2019) International Atomic Energy Agency, IAEA Environment Laboratories, Monaco (Monaco). Environ Laboratories Newsl, 6(1)

IAEA (2020) Nuclear power and the clean energy transition. IAEA Bulletin

IAEA (2021a) Nuclear science for ocean health. https://www.iaea.org/newscenter/multimedia/videos/nuclear-science-for-ocean-health. Accessed 13 Jan 2025

IAEA (2021b) Nuclear Technology for Controlling Plastic Pollution (NUTEC). https://www.iaea.org/sites/default/files/21/05/nuclear-technology-for-controlling-plastic-pollution.pdf. Accessed 15 Jan 2025

Kelleway JJ, Saintilan N, Macreadie PI, Baldock JA, Heijnis H, Zawadzki A et al (2017) Geochemical analyses reveal the importance of environmental history for blue carbon sequestration. J Geophys Res Biogeosci 122(7):1789–1805

Khodadadi M, Gibbs M, Swales A, Toloza A, Blake WH (2024) Anthropogenic and climatic impacts on historic sediment, carbon, and phosphorus accumulation rates using $^{210}Pb_{ex}$ and ^{137}Cs in a sub-watershed linked to Zarivar Lake, Iran. Environ Monit Assess 196(10):887

Klubi E, Abril J, Nyarko E, Laissaoui A, Benmansour M (2017) Radioecological assessment and radiometric dating of sediment cores from dynamic sedimentary systems of Pra and Volta estuaries (Ghana) along the Equatorial Atlantic. J Environ Radioact 178–179:116–126

Kuznetsova OV, Shtykov SN, Timerbaev AR (2023) Mass spectrometry insight for assessing the destiny of plastics in seawater. Polymers 15(6):1523

Laissaoui A, Mejjad N, Ziad N, Ait Bouh H, El Hammoumi O, Benkdad A, Fekri A (2018) Evidence for a recent increase in delivery of atmospheric ^{210}Pb to Oualidia lagoon, coastal Morocco. Environ Monit Assess 190(11):642
Lamont KM (2022) Temporal carbon assessment through radiometric and repeat measure analysis in a mangrove. Doctoral dissertation, Macquarie University
Lanctôt CM, Al-Sid-Cheikh M, Catarino AI, Cresswell T, Danis B, Karapanagioti HK, Mincer T, Oberhänsli F, Swarzenski P, Tolosa I, Metian M (2018) Application of nuclear techniques to environmental plastics research. J Environ Radioact 192:368–375
Lawson IT, Honorio Coronado EN, Andueza L, Cole L, Dargie GC, Davies AL et al (2022) The vulnerability of tropical peatlands to oil and gas exploration and extraction. Prog Environ Geogr 1(1–4):84–114
Levin LA, Hönisch B, Frieder CA (2015) Geochemical proxies for estimating faunal exposure to ocean acidification. Oceanography 28(2):62–73
Lind OC, Stegnar P, Tolongutov B, Rosseland BO, Strømman G, Uralbekov B et al (2013) Environmental impact assessment of radionuclide and metal contamination at the former U site at Kadji Sai, Kyrgyzstan. J Environ Radioact 123:37–49
Ling SD, Johnson CR, Frusher SD, Ridgway K (2009) Overfishing reduces resilience of kelp beds to climate-driven catastrophic phase shift. Proc Natl Acad Sci 106(52):22341–22345
Maher DT, Santos IR, Schulz KG, Call M, Jacobsen GE, Sanders CJ (2017) Blue carbon oxidation revealed by radiogenic and stable isotopes in a mangrove system. Geophys Res Lett 44(10):4889–4896
Mein TF, Gimenes ALV, Dias EM, Rebello Pinho Dias Scoton ML, Udaeta MEM (2021) Environmental vulnerability in pre-salt oil and gas operations. Energies 14(3):732
Mejjad N, Laissaoui A, El-Hammoumi O, Benmansour M, Benbrahim S, Bounouira H et al (2016) Sediment geochronology and geochemical behavior of major and rare earth elements in the Oualidia Lagoon in the western Morocco. J Radioanal Nucl Chem 309(3):1133–1143
Mejjad N, Laissaoui A, El-Hammoumi O, Fekri A, Amsil H, El-Yahyaoui A, Benkdad A (2018) Geochemical, radiometric, and environmental approaches for the assessment of the intensity and chronology of metal contamination in the sediment cores from Oualidia lagoon (Morocco). Environ Sci Pollut Res 25:22872–22888
Mejjad N, Laissaoui A, Fekri A, Hassen NEH, Benmhammed A, El Hammoumi O et al (2022) Tracking natural and human impact on sediment dynamics using radiometric approach in Oualidia lagoon (Morocco). Int J Environ Anal Chem 102(16):4300–4315
Mejjad N, Laissaoui A, El Mansouri B, Fekri A, Moumen A, El Khalidi K, El Hammoumi O (2023) Bibliometric analysis of the literature on coastal sediment pollution. In: Coasts, estuaries and lakes. Springer, pp 3–24
Meybeck M, Lestel L, Bonté P, Moilleron R, Colin JL, Rousselot O et al (2007) Historical perspective of heavy metals contamination (Cd, Cr, Cu, Hg, Pb, Zn) in the Seine River basin (France) following a DPSIR approach (1950–2005). Sci Total Environ 375(1–3):204–231
Munir M, Sholikhah UN, Lestari E, Pujiyanto A, Prasetya KE, Nurmanjaya A, Sarwono DA, Subechi M, Suseno H (2023) Iodine-131 radiolabeled polyvinylchloride: a potential radiotracer for micro and nanoplastics bioaccumulation and biodistribution study in organisms. Mar Pollut Bull 188:114627
Musat F, Vogt C, Richnow HH (2016) Carbon and hydrogen stable isotope fractionation associated with the aerobic and anaerobic degradation of saturated and alkylated aromatic hydrocarbons. J Mol Microbiol Biotechnol 26(1–3):211–226
Patterson III WF, Chanton JP, Hollander DJ, Goddard EA, Barnett BK, Tarnecki JH (2020) The utility of stable and radioisotopes in fish tissues as biogeochemical tracers of marine oil spill food web effects. In: Scenarios and responses to future deep oil spills: fighting the next war. Springer, pp 219–238
Piazza G (2023) Geochemical and mineralogical studies to assess the significance of temperate coralline algae as paleoclimate archives. Tesi di dottorato, Università degli Studi di Milano-Bicocca

Ponomarev A, Gohs U, Ratnam T, Horak C (2022) Keystone and stumbling blocks in the use of ionizing radiation for recycling plastics. Radiat Phys Chem 201:110397
Prinzhofer A, Battani A (2003) Gas isotopes tracing: an important tool for hydrocarbons exploration. Oil Gas Sci Technol 58(2):299–311
Pyl M, Taylor A, Oberhänsli F, Swarzenski P, Besson M, Danis B, Metian M (2022) Evidence of microplastic-mediated transfer of PCB-153 to sea urchin tissues using radiotracers. Mar Pollut Bull 185:114322
Qiao J, Casacuberta N, Ginnity PM (2023) Editorial: Natural and artificial radionuclides as tracers of ocean processes. Front Mar Sci 10:1170408
Rabinovitch L, Saturos G, McCallum P, Kwok H, Yan J, Filewood T, Cody R, Brunswick P, Shang D (2024) Rapid analysis of spilled petroleum oils by direct analysis in real time time-of-flight mass spectrometry with hydrophobic paper sample collection. Environ Syst Res 13(1):1–16
Rafa N, Ahmed B, Zohora F, Bakya J, Ahmed S, Ahmed SF, Mofijur M, Chowdhury AA, Almomani F (2024) Microplastics as carriers of toxic pollutants: source, transport, and toxicological effects. Environ Pollut 343:123190
Rogers K, Kelleway JJ, Saintilan N (2023) The present, past and future of blue carbon. In: Cambridge prisms: coastal futures, vol 1. Cambridge University Press, p e30
Sefton J, Woodroffe S, Ascough P (2020) Radiocarbon dating of mangrove sediments. In: Dynamic sedimentary environments of mangrove coasts, pp 199–215
Sirin M (2019) Investigation of accumulation of radionuclides in different tissues of Whiting fish (Merlangius merlangus euxinus Nordmann, 1840) caught on the coasts of Rize in the eastern Black Sea region of Turkey. Microchem J 152:104349
Smith M, Love DC, Rochman CM, Neff RA (2018) Microplastics in seafood and the implications for human health. Curr Environ Health Rep 5(3):375
Sumaila UR, Tai TC (2020) End overfishing and increase the resilience of the ocean to climate change. Front Mar Sci 7:523
Taieb Errahmani D, Noureddine A, Abril-Hernández JM, Boulahdid M (2020) Environmental radioactivity in a sediment core from Algiers Bay: radioecological assessment, radiometric dating and pollution records. Quat Geochronol 56:101049
Taipale SJ, Peltomaa E, Kukkonen JV, Kainz MJ, Kautonen P, Tiirola M (2019) Tracing the fate of microplastic carbon in the aquatic food web by compound-specific isotope analysis. Sci Rep 9(1):1–15
Tang D, Luo S, Deng S, Huang R, Chen B, Deng Z (2022) Heavy metal pollution status and deposition history of mangrove sediments in Zhanjiang Bay, China. Front Mar Sci 9:989584
Torres FG, De-la-Torre GE (2021) Historical microplastic records in marine sediments: current progress and methodological evaluation. Reg Stud Mar Sci 46:101868
Trajano JCI (2021) Battling marine plastic waste: nuclear technology's role. RSIS Commentaries 019-21. RSIS Publications
Turner S, Horton AA, Rose NL, Hall C (2019) A temporal sediment record of microplastics in an urban lake, London, UK. J Paleolimnol 61(4):449–462
Uddin S, Fowler SW, Uddin MF, Behbehani M, Naji A (2021) A review of microplastic distribution in sediment profiles. Mar Pollut Bull 163:111973
Van Eck NJ, Waltman L (2022) VOSviewer software version 1.6.18. Leiden University's Centre for Science and Technology Studies (CWTS)
Vidal A, Seignemartin G, Copard Y, Montargès-Pelletier E, Ollive V, Papillon L, Grenz C, Eyrolle F, Sempéré R (2024) Temporal trends of plastic additive contents in sediment cores of three French rivers (Loire, Meuse and Moselle) over the last decades. Sci Total Environ 931:172849
Vieth A, Wilkes H (2010) Stable isotopes in understanding origin and degradation processes of petroleum. In: Timmis KN (ed) Handbook of hydrocarbon and lipid microbiology. Springer, Berlin, pp 87–120
Wang YS, Gu JD (2021) Ecological responses, adaptation and mechanisms of mangrove wetland ecosystem to global climate change and anthropogenic activities. Int Biodeterior Biodegrad 162:105248

Wang Z, Stout SA, Fingas M (2006) Forensic fingerprinting of biomarkers for oil spill characterization and source identification. Environ Forensic 7(2):105–146
Wang B, Hua L, Mei H, Wu X, Kang Y, Zhao N (2024) Impact of climate change on the dynamic processes of marine environment and feedback mechanisms: an overview. Arch Comput Methods Eng:1–32
Wei Y, An M, Huang J, Fang X, Song M, Wang B et al (2024) How human activities affect and reduce ecological sensitivity under climate change: case study of the Yangtze River economic belt, China. J Clean Prod 472:143438
Whitehead PG, Wilby RL, Battarbee RW, Kernan M, Wade AJ (2009) A review of the potential impacts of climate change on surface water quality. Hydrol Sci J 54(1):101–123
Williamson P, Gattuso JP (2022) Carbon removal using coastal blue carbon ecosystems is uncertain and unreliable, with questionable climatic cost-effectiveness. Front Clim 4:853666
Wu F, Zonneveld KA, Wolschke H, von Elm R, Primpke S, Versteegh GJ, Gerdts G (2024) Diving into the depths: uncovering microplastics in Norwegian coastal sediment cores. Environ Sci Technol 58(38):17036–17047
Xiao J, Xiao YK, Jin ZD, He MY, Liu CQ (2013) Boron isotope variations and its geochemical application in nature. Aust J Earth Sci 60(4):431–447
Yang C, Wang Z, Hollebone BP, Brown CE, Yang Z, Landriault M (2014) Chromatographic fingerprinting analysis of crude oils and petroleum products. In: Handbook of oil spill science and technology. Wiley, pp 93–163
Yang S, Yang Q, Song X, Zhou W, Liu J, Qi X, Chen J, Huang J, Li B, Sun Y (2024) Blue carbon in sediment from Sanggou Bay: composition, burial flux and its response to human activities. Front Mar Sci 10:1342750
Zhang H, Shan B (2008) Historical records of heavy metal accumulation in sediments and the relationship with agricultural intensification in the Yangtze–Huaihe region, China. Sci Total Environ 399(1–3):113–120
Zhang L, Du F, Wang Y, Li Y, Yang C, Li S, Huang X, Wang C (2021) Oil fingerprint identification technology using a simplified set of biomarkers selected based on principal component difference. Can J Chem Eng 100(1):23–34
Zhou Z, Wang Y, Teng H, Yang H, Liu A, Li M, Niu X (2020) Historical evolution of sources and pollution levels of heavy metals in the sediment of the Shuanglong Reservoir, China. Water 12(7):1855

Chapter 2
An Analysis of the Application of Nuclear and Isotopic Approaches for the Monitoring of Organic and Inorganic Pollutants

Souad Nasrdine, Nezha Mejjad, Jamal Mabrouki, Youssef El Mourabit, and Miloudia Slaoui

Abstract Marine and coastal environments are indispensable for biodiversity and climate change mitigation, threatening by anthropogenic pollutants such as radionuclides, heavy metals, microplastics among other organic and inorganic pollutants. Advanced and innovative approaches and methodologies are required to assess and trace pollutants origins, quantify and evaluate impacts, and inform decision making concerning sustainable mitigation and management strategies. The present chapter analyze the critical role of nuclear and isotopic methods in addressing organic and inorganic pollution in coastal environments. The analysis revealed the role of radionuclides such as ^{137}Cs and ^{210}Pb in reconstructing pollution history, identifying trends and potential sources to linking contamination significant levels to specific anthropogenic activities. Stable isotopes ($\delta^{13}C$, $\delta^{15}N$) enable pollutants origins distinction and differentiating agricultural runoff from industrial discharges while emerging applications involve stable isotope fingerprinting of microplastics to trace the sources of polymer and isotopic tracers (e.g., ^{3}H, ^{129}I) for modeling dispersion of pollutant in ocean currents. Through bridging scientific innovation with policy

S. Nasrdine (✉) · M. Slaoui
Energy, Materials and Sustainable Development (EMDD) Laboratory—Higher School of Technology—SALE, Center of Water, Natural Resources, Environment and Sustainable Development (CERN2D), University Mohammed V in Rabat, Rabat, Morocco

N. Mejjad
National Centre for Nuclear Energy, Science and Technology (CNESTEN), Rabat, Morocco

J. Mabrouki
Laboratory of Spectroscopy, Molecular Modeling, Materials, Nanomaterials, Water and Environment, CERNE2D, Faculty of Science, Mohammed V University in Rabat, Rabat, Morocco

Y. El Mourabit
Aquatic Systems: Marine and Continental Ecosystems Lab, University of Ibn Zohr, Agadir, Morocco

N. Mejjad (ed.), *Nuclear-Based Approaches application for marine pollution assessment and monitoring*, Environmental Science and Engineering,
https://doi.org/10.1007/978-3-032-12197-4_2

frameworks, nuclear and isotopic techniques provide a roadmap for mitigating marine pollution, increasing ecosystem resilience, and safeguarding life below water and thus human health.

Keywords Marine pollution · Nuclear techniques · Isotopic tracers · Radiotracers · Radionuclides · Isotope analysis

2.1 Introduction

Oceans are considered as the foundation of life on Earth, as they paly essential role in regulating the climate, producing oxygen in the atmosphere, and preserving biodiversity. They are also a vital resource for humanity, by providing food, electricity, and livelihoods for hundreds of millions of people. Nevertheless, this essential ecosystem is under unprecedented stress, resulting from man-made factors and environmental disturbances (Sandifer and Sutton-Grier 2014; Stoeckl et al. 2024; Mejjad et al. 2025). Nevertheless, this essential ecosystem is under unprecedented stress, resulting from anthropogenic factors and environmental disturbances. Marine and coastal ecosystems are under serious threat from a range of pollutants. These include toxic substances such as heavy metals, hydrocarbons, microplastics and persistent chemicals (Marchant et al. 2016), which enter the oceans from both terrestrial and marine sources. This pollution is mainly due to industrial, agricultural and urban activities, as well as maritime practices. In addition, the effects of climate change, such as rising of ocean temperatures, acidification and sea-level rise, are altering the dynamics of contaminant diffusion and impact (Wang et al. 2021).

To remedy this environmental threat, it has become imperative to develop reliable tools to assess the state of health of marine ecosystems and to understand the propagation mechanisms and impacts of pollutants. Indeed, among the available approaches, nuclear and isotopic techniques stand out for their accuracy and ability to trace the origin of pollutants, track their spread and quantify their impact on marine ecosystems. These techniques provide a means of solving the problems posed by marine pollution, focusing on pollutants, such as microplastics and nanomaterials, and which require advanced analytical methods (Mancinelli and Vizzini 2015).

This chapter takes an in-depth look at the use of nuclear and isotopic techniques to assess marine pollutants. It highlights their application in the monitoring of organic and inorganic pollutants, their propagation in marine ecosystems and their role in sustainable ocean management. It proposes a global vision for the protection of marine ecosystems based on the challenges and opportunities offered by these technologies.

2.2 Human Pressures on Marine Ecosystems

2.2.1 Overview of Pollutants Affecting Marine and Coastal Ecosystems

Marine and coastal ecosystems have been increasingly endangered by various sources of man-made pollution. Whether brought in directly or indirectly by human activities, these pollutants have a significant effect on the health and sustainability of marine environments (Fig. 2.1). Activities such as agriculture and related practices around marine and coastal systems release various type of pollutants including heavy metals (e.g., Cd, Cr, Cu), plastics, exhaust fumes, methane among other types (Thiagarajan and Devarajan 2024). Industrial activities generate in atmosphere, coastal and river systems different types of pollution which reach ocean through precipitation in case of pollutants released in air while it flows downstream through rivers (Mishra et al. 2023; Tavengwa et al. 2021). Other activities practiced in the ocean are affecting it and generating various types of pollutants such as fishing, maritime transport, offshore mining, and coastal tourism (Mejjad et al. 2022; Walker et al. 2019) (Fig. 2.1).

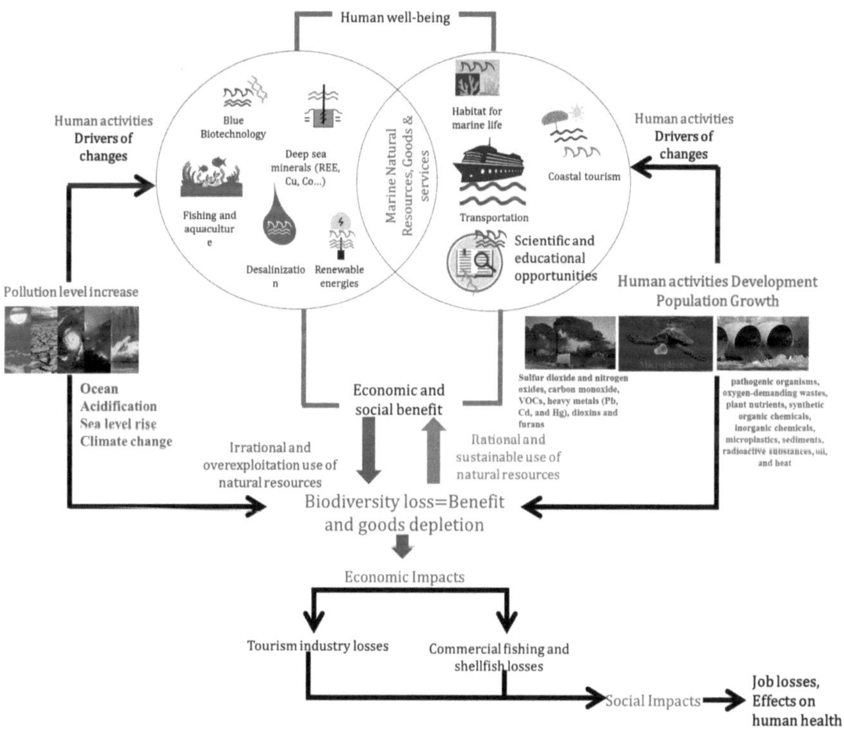

Fig. 2.1 Different human activities affecting ocean environment and main released pollution and potential impacts. (Designed by ***N. Mejjad***)

The main types of pollutants generated by human activities are as follow:

Organic pollutants. Organic pollutants can cause severe human diseases if their concentrations exceed the permissible limits. They are toxic molecular compounds that have as origins industrial products such as petroleum hydrocarbons, detergents, organic solvents, plastics, pesticides, and dyes (Ghosh et al. 2020). These products, when introduced into the marine environment, can accumulate in food chains and have toxic effects on marine organisms and thus on human health. This can affect in turn human well-being where biodiversity loss can influence both economic and social sectors (Fig. 2.1).

Heavy metals. Heavy metals were since decades among the most discussed and investigated pollutants where several research projects were carried out for the assessment and monitoring of heavy metals pollution degree and mitigations technologies (Wang et al. 2022; Wan et al. 2016; Marın et al. 2001). The assessment of heavy metals (HM) concentrations and identification of their potential sources is the first step toward developing management tools and technologies. Among the used approaches that allowed for defining the heavy metals sources is the nuclear methods, whereby establishing relationship age-depth using radiotracers such as ^{210}Pb, and ^{137}Cs it is possible to track changes of HM concentrations in sediment core (Laissaoui et al. 2018) over time. Thus, by investigating history of natural and anthropogenic activities occurred in the studied zone it's possible to define potential source of enrichment by HM in the zone under investigation.

Heavy metals can have as origin industrial activities, agricultural practices and domestic activities. HM such as lead, mercury, chromium and cadmium resulting from these activities present a major threat to coastal and marine systems and the biodiversity (Thiagarajan and Devarajan 2024; Mok et al. 2023; Pandiyan et al. 2021; Shah 2021). These substances are non-biodegradable and can accumulate in biological tissues, disrupting the physiological functions of marine species (Sable et al. 2024; Shahjahan et al. 2022; Stankovic et al. 2014).

Radionuclides. The global oceans are a geochemical well of various pollutants, including radionuclides, which come from both natural and anthropogenic sources (Figs. 2.2 and 2.3). The latter mainly originates from fallout from past atmospheric nuclear weapons tests and releases following accidents at nuclear facilities.

Radionuclides, also known as radioisotopes, are unstable chemical species that emit excess energy through spontaneous radioactive disintegration (Choppin et al. 2013). In pollution monitoring, detection includes alpha particles (helium nuclei), beta particles (electrons), and gamma photons across an energy spectrum from tens of keV (kilo electron volts) to MeV (megaelectron volts).

Natural radionuclides have three sources, each of which contributes to the Earth's background radiation and influences radiation exposure worldwide (Fig. 2.3). These three sources are, primordial radionuclides (e.g., Uranium-238 (^{238}U), Thorium-232 (^{232}Th), and Potassium-40 (^{40}K)), cosmogenic radionuclides

Natural radionuclides

Found throughout the environment (Air, soil, water).

Artificial radionuclides

Released by human activity

Categories

Natural radionuclides:

1. **Radionuclides** induced in Earth's atmosphere by **cosmic rays** (e.g. H-3, Be-10, C-14).
2. **Non chain primordial radionuclides** (e.g. K-40, Rb-87). Activity per unit mass of K-40 is measured in building materials for radiation protection purposes.
3. **Uranium-238** chain radionuclides (e.g. Ra-226, Rn-222, Pb-214, Bi-214, Pb-210). Activity per unit mass of Ra-226 is measured in building materials for radiation protection purposes. Concentration of Rn-222 is measured in soil and indoor air for radiation protection purposes.
4. **Thorium-232** chain radionuclides (e.g. Th-228, Rn-220). Activity per unit mass of Th-228 is measured in building materials for radiation protection purposes. Concentration of Rn-220 is measured in indoor air as a complement to Rn-222 concentration.
5. **Uranium-235** chain radionuclides (e.g. Th-231, Pa-231).

Artificial radionuclides:

1. **C-14, Ar-41,** iodine radioisotopes, xenon radioisotopes, krypton radioisotopes in air.
2. **H-3, Co-60, Sr-90, Cs-134, Cs-137** in water.
3. **Sr-90, Cs-134, Cs-137,** uranium radioisotopes, plutonium radioisotopes, americium radioisotopes, curium radioisotopes in soil.

Potential sources

1. **Nuclear accidents** in nuclear facilities and the spread of radionuclides into the environment.
2. **Radiological incidents** in workplaces with radioactive substances.
3. Insufficient measures against the **penetration of wastes** into the environment from waste repositories.
4. The use of **nuclear weapons.**

Fig. 2.2 Detailed description of radionuclide categories

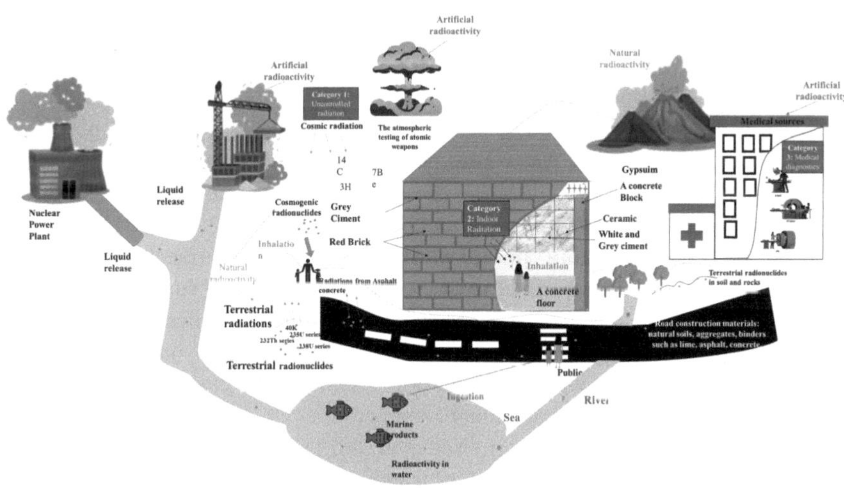

Fig. 2.3 Natural and anthropogenic sources of radionuclides. (Designed by *N. Mejjad*)

(Carbon-14 (^{14}C), Tritium (^{3}H), and Beryllium-10 (^{10}Be)), and decay products (Radon-222 (^{222}Rn)) (Choppin et al. 2013; Eisenbud and Gesell 1997).

Anthropogenic radionuclides arise from human-made activities that use or produce radioactive isotopes (Fig. 2.3). Reactors and nuclear power plants contribute to the generation and production of radionuclides. The IAEA report on radiation protection standards reports that nuclear reactors contribute to radionuclide generation through the processes of cooling water and waste (IAEA 2024) while via fission and neutron activation, nuclear power generation produces radionuclides (Eisenbud and Gesell 1997). Also, nuclear weapons testing has severely contributed to the generation of radionuclides such as Cesium-137 (^{137}Cs) and Plutonium-239 (^{239}Pu), mainly between the period extending from 1940 to 1980. Based on the Committee on the Effects of Atomic Radiation, nuclear weapons tests are the main source of environmental contamination by radionuclides (UNSCEAR, 2000 Report, 2000, Annex C).

Industrial and Medical Applications are also contributing to radionuclides production and release into the environment, where Technetium-99m (^{99m}Tc) is used for medical diagnostics and Iridium-192 (^{192}Ir) for industrial radiography (Choppin et al. 2013). Nuclear fuel cycle activities, such as milling, uranium mining, and fuel reprocessing, generate radionuclides such as Uranium-238 (^{238}U) and Radium-226 (^{226}Ra), and they are identified as a major source of radionuclides (UNSCEAR, 2014). Incidents and nuclear accidents have released artificial radionuclides such as Cesium-137 (^{137}Cs) and Iodine-131 (^{131}I) on a large scale such as the accident of Fukushima (2011) and Chernobyl (1986). According to UNSCEAR (2000), these nuclear events have marked these periods and constitute a major anthropogenic sources. In addition, IAEA indicate that even the smaller incidents at waste facilities contribute in radionuclides release at local level (IAEA, Radiation Protection Standards, 2014). Also, it is worth noting that inadequate disposal of radioactive waste lead to radionuclides release such as ^{137}Cs into environmental components (e.g. soil, water) (Choppin et al. 2013). Accordingly, radioactive waste must be managed in an adequate manner to avoid any potential release of radioactive elements into natural environments and thus prevent any effects on biodiversity and human health.

Microplastics and emerging contaminants. Plastics waste can easily reach coastal and marine environments because of the mismanagement of such waste (Mejjad et al. 2023). The degradation of plastic waste results on microplastics among others emerging pollutants such as innovative chemicals or nanomaterials. Plastics can break down into microplastics (MP) and nanoplastics (NP) when exposed to a mixture of fragmentation processes called physical, chemical and biological degradations (Swee-Li Yee et al. 2021) (Fig. 2.4). These MP and NP can contain plasticizers and additives leading to pollution with emerging pollutants known as innovative chemicals and nanomaterials.

Based on studies carried out by De-la-Torre and Aragaw (2021; Pan 2023; Soto-Navarro et al. 2021) 80% of pollutants entering ocean are from land based sources, 10% from ocean sources while plastic and microplastics account for 10%, but their persistence in the environment and their capacity to accumulate in the food chain make them of particular concern.

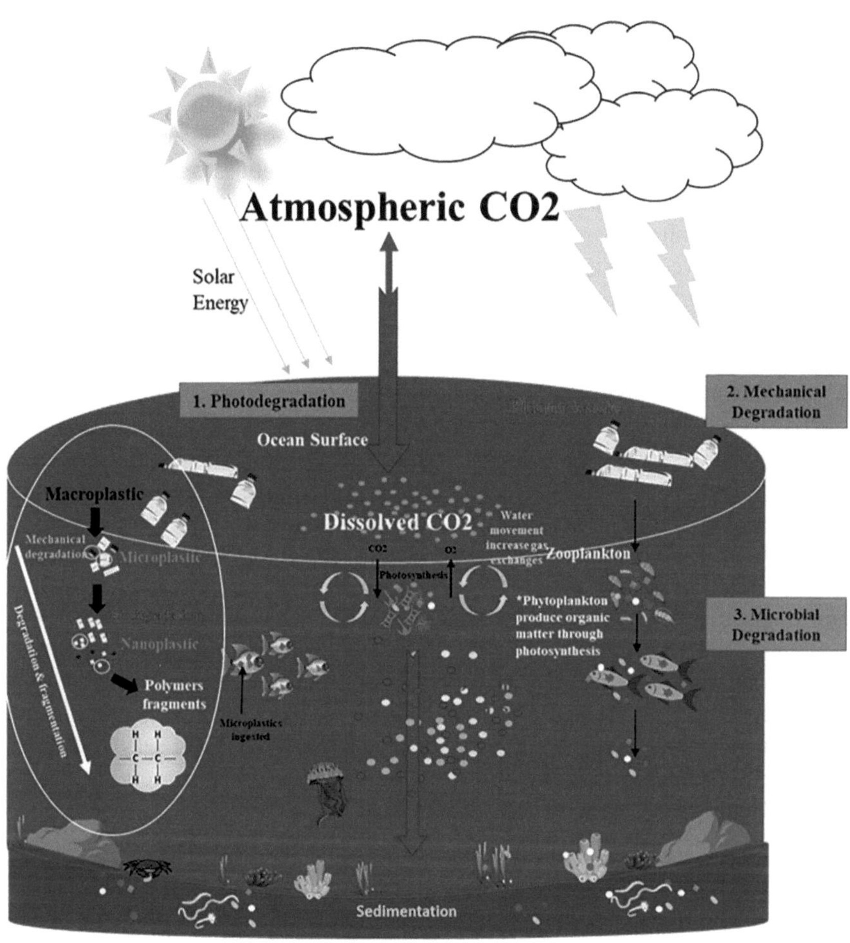

Fig. 2.4 Microplastics degradation and fragmentation. (Designed by *N. Mejjad*)

2.3 Radionuclides Applications in Marine Pollution Studies

Nuclear and isotopic approaches (NIA) were largely utilized for pollution levels assessment (IAEA 2024, 2025). Since decades, nuclear and isotopic techniques were employed as tool for pollution monitoring in coastal and marine pollution (Aarkrog 1998; Livingston and Povinec 2000; Laissaoui et al. 2008, 2013; Benkdad et al. 2008, 2011; Omokheyeke et al. 2014; Mejjad et al. 2018, 2022). The use of NIA can help in decision making where for example stable isotopes such as $\delta^{13}C$, $\delta^{15}N$, $\delta^{34}S$ allow identifying sources of marine pollutants. Based on Kendall et al. (2007), the analysis of $\delta^{15}N$ allows for the distinction between nitrogen inputs from sewage discharges, agricultural runoff, or industrial effluents in coastal waters, enabling policymakers to develop specific controls on the main pollution source.

Radionuclides are used as valuable tools for monitoring marine pollution. The application fields include (Table 2.1):

Tracing the pollutant sources. Plutonium-239/240 ($^{239+240}Pu$), cesium-137 (^{137}Cs), and lead-210 (^{210}Pb) aid in defining the sources of contaminants. For example, ^{137}Cs is used as a radiotracer. These artificial radionuclides primarily originated from nuclear testing and accidents such as Chernobyl and Fukushima, can trace radioactive pollution spread (Ory et al. 2021). Anthropogenic radionuclides distinguish between natural background levels and human-induced pollution.

Establishment of pollution history. Radionuclides such as ^{210}Pb and ^{137}Cs can reconstruct the history of organic and inorganic pollutants' deposition in coastal and marine environments. The radiometric dating of sediment cores enables the establishment of the history of pollution (Laissaoui et al. 2018; Mejjad et al. 2016). Pb-210 is ideal for dating sediments over the last 100–150 years, identifying temporal changes in pollutant levels such as heavy metals (Ferrand et al. 2012; Ligero et al. 2002), rare earth elements (Mejjad et al. 2022), microplastics (Tsabaris et al. 2024), among other types of pollutants. ^{137}Cs is an independent radiotracer used to trace specific events such as the 1986 Chernobyl fallout or the 1963 nuclear testing peak.

Tracking water mass movement (Kaufman and Orlob 1956; Noureddine et al. 2008): Tritium (^{3}H) and Carbon-14 (^{14}C) can be employed for tracing ocean currents and water mass mixing. Also, tracers released from local sources such as Cap de la Hague-derived ^{129}I and Sellafield-derived ^{137}Cs and ^{90}Sr were considerably used to trace water transport in the Arctic ocean, North and Baltic Seas (Livingston 1988; Kershaw and Baxter 1993). Thus, using radionuclides tracers such as ^{14}C and ^{3}H allows tracing water mass mixing and ocean current which enable model

Table 2.1 Key radionuclides in marine pollution studies

Radionuclide	Half-life	Source	Key applications
^{137}Cs	30.2 years	Nuclear accidents, weapons testing	Track ocean currents, sediment mixing, bioaccumulation in fish
$^{239}Pu/^{240}Pu$	24,100/6560 years	Nuclear weapons fallout, waste	Long-term sediment chronology, legacy pollution (e.g., Arctic dumping)
^{90}Sr	28.8 years	Nuclear reactors, accidents	Bone tissue bioaccumulation in marine mammals
^{210}Pb	22.3 years	Natural (uranium decay)	Sediment dating (100–150 years), erosion studies
^{234}Th	24.1 days	Natural (uranium decay)	Short-term particle scavenging near pollution sources
^{3}H (Tritium)	12.3 years	Cosmogenic, nuclear reactors	Trace recent water movement, groundwater discharge
^{14}C	5730 years	Cosmic rays, nuclear testing	Distinguish fossil fuel vs. natural CO_2 in ocean acidification
^{129}I	15.7 million years	Nuclear reprocessing, accidents	Ultra-long-term tracer for nuclear waste leaks, ocean circulation (decadal-millennial scale)

pollutants dispersion such as microplastics, oil or chemical effluents, enabling the prediction of pollutants pathways and areas of accumulation.

Bioaccumulation studies. The occurrence of radionuclides in marine organisms such as mollusks and fish indicate pollutants bioaccumulation (Tan et al. 2023; Risal et al. 2024). Measuring activity concentrations of ^{137}Cs or ^{210}Po in biota allows assessing the radioactive and associated non-radioactive pollutants transfer through food webs, reflecting ecosystem health (Santhanabharathi et al. 2025).

Isotopes and radioactive tracers play an essential role in advancing the understanding of oceanic processes where by merging stable isotopes where by merging stable isotope hydrology with radiometric dating, oceanographers unravel the pathways and timescales of water masses, enhancing our understanding of biogeochemical cycles, climate dynamics, and marine ecosystem connectivity.

2.4 Nuclear Approaches and Methods Used for Pollutant Measurement

The use of nuclear and isotopic methods for inorganic and organic pollutants helps in pollution mitigation, policy decisions, and ecological risk assessment. Figure 2.5 shows different techniques employed for organic and inorganic pollutants measurement, applications, and common workflow and outcomes.

2.4.1 Inorganic Pollutants Analysis

Neutron activation analysis (NAA). Neutron activation analysis (NAA) is used to measure and quantify heavy metals such as arsenic, cadmium, mercury, and rare earth elements (REEs) in environmental matrices such as sediment (Mejjad et al. 2018, 2022) soil (El-Taher and Abdelhalim 2014; Zaim et al. 2016), water (Tanizaki 1990; Greenberg and Kingston 1983; Blanchard et al. 1959) and biological samples (Antsiferova et al. 2015; Das et al. 2024; Ohta et al. 1996) NAA is among the most sophisticated techniques for trace metals and is known as a non-destructive analytical method. The method principle is based on the irradiation of samples with neutrons. The samples become radioactive and emit gamma rays with unique energy signatures, allowing precise identification of elements (metals and rare earth elements) at trace levels (ppb to ppt) (Witkowska et al. 2005).

By analyzing pollutant levels in environmental and biological samples, NAA allows for identifying pollution sources, such as industrial discharge vs. natural geogenic arsenic in groundwater. Through the quantification of toxic metals in industrial effluent, NAA monitors compliance with environmental regulations. In addition, pinpointing hotspot areas of heavy metals, mainly near mining sites, helps guide soil remediation.

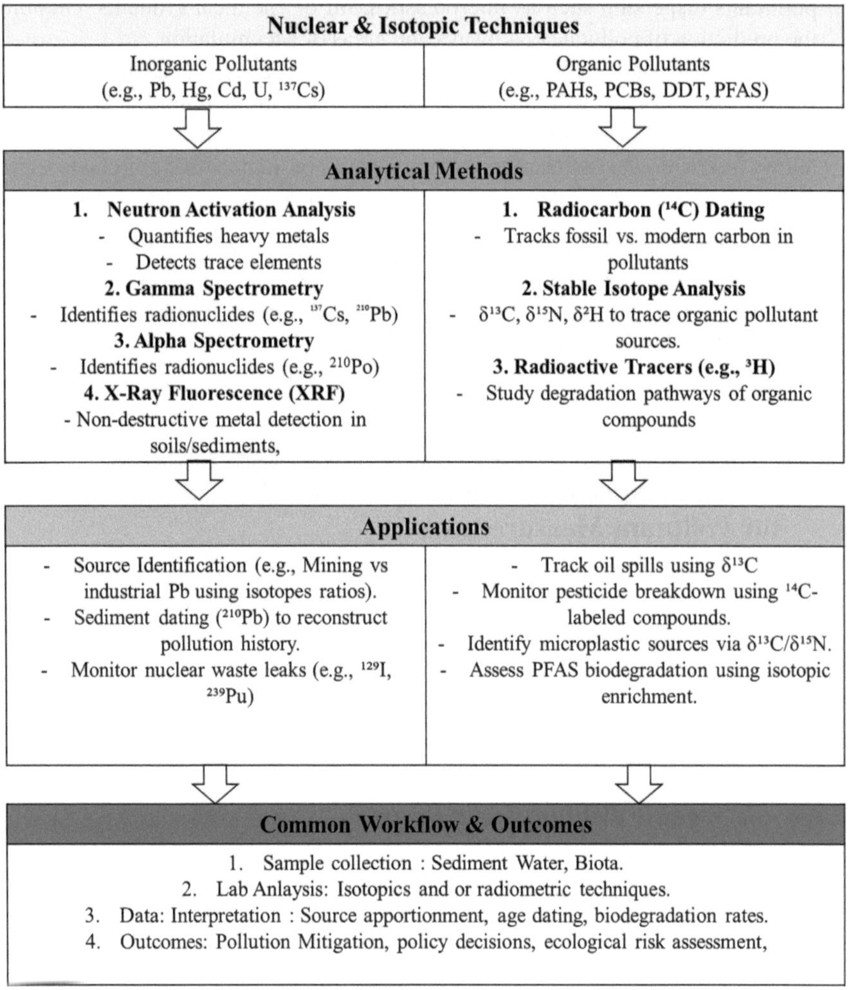

Fig. 2.5 Overview of nuclear and isotopic techniques for water and environmental analysis

Gamma spectrometry. This technique detects and quantifies radioactive elements emitting gamma rays, such as Ra-226, Pb-210, and Cs-137. Gamma spectrometry is essential for sediment dating (see Sect. 2.3), which allows informing restoration and management strategies by reconstructing the history of pollution trends over the last 100–150 years. In addition, the measure of activity concentration of natural and artificial radionuclides such as Cs-137 allowing mapping their dispersion from nuclear accident and related events (e.g., Fukushima) and thus guiding coastal cleanup and seafood safety protocols.

Alpha spectrometry. This nuclear technique involves the measurement of alpha particles, which are helium nuclei emitted during the radioactive decay of certain nuclides (Aggarwal 2016). This analytical technique requires radiochemistry preparation and uses semiconductor detectors, such as silicon surface barrier detectors, to achieve high energy resolution, enabling the identification and quantification of specific alpha-emitting radionuclides like Pu-240, Pu-239, Po-210, U-238, and Am-241. Alpha spectrometry is essential for environmental monitoring, and it is widely used in research program projects as it is a highly sensitive method and enables detection of low contamination levels in complex matrices such as sediment, seawater, and soil. It is used to analyze marine organisms, sediment, and seawater, helping in assessing pollution degree from different sources such as industrial discharges or nuclear accidents, providing data on pollutant concentrations and type to assess potential risks to marine life, biodiversity, and human health (Jia and Jia 2012; Bisinger et al. 2010; Zaborska et al. 2007).

Ray fluorescence (XRF). X-Ray Fluorescence (XRF) is known as a non-destructive technique that measures and identifies inorganic pollutants such as chromium (Cr), lead (Pb) in marine sediment, soils, and industrial waste by measuring emitted X-rays (Oladebeye 2017; West et al. 2015).

2.4.2 *Organic Pollutants Analysis*

Radiocarbon (^{14}C) Dating. Using ^{14}C allows for the distinction between modern organic matter and fossil-derived pollutants such as coal, petroleum hydrocarbons, and PAHs. Due to their ancient origin, fossil fuels lack ^{14}C, while it is retained by modern organic material (Norris 2015; Lichtfouse et al. 1997). ^{14}C is essential to track oil spills to specific sources, such as differentiating between natural seeps and pipeline leaks (Conrad et al. 1997). Also, the radiocarbon isotope method is reported as a powerful tool for tracking carbon emissions (Zhang et al. 2021) as well as sources of pollutants in air (Gustafsson et al. 2009; Szidat et al. 2004), which helps in decision making regarding the mitigation of climate change effects.

Stable Isotope Analysis ($\delta^{13}C$, $\delta^{15}N$, $\delta^{2}H$) using Elemental Analyzer coupled to Isotope ratio mass spectrometry (EA-IRMS). Stable isotopes such as $\delta^{13}C$, $\delta^{15}N$, and $\delta^{2}H$ act as fingerprints to trace organic pollutants such as microplastics, pesticides, and PFAS (per- and poly-fluoroalkyl substances) (Schwarzbauer and Jovančićević 2024). They are measured using Isotope ratio mass spectrometry (IRMS) analyzers. IRMS is a Mass spectrometry method for measuring the relative stable isotopic abundances of the elements in a material (Perini and Bontempo 2022). Stable isotope ratios of carbon ($^{13}C/^{12}C$), oxygen ($^{18}O/^{16}O$), hydrogen ($^{2}H/^{1}H$), nitrogen ($^{15}N/^{14}N$), and sulfur ($^{34}S/^{32}S$) are measured to provide data and information about the chemical, geographic and biological origins of substances (Scrimgeour and Robinson 2003; Li 2011). When coupled to EA, IRMS-EA is used as a powerful tool for characterizing polymers of plastics in the marine environment (Barbieri et al. 2023; Berto et al. 2019). These stable isotopes allow the distinction between industrial and agricultural sources of pesticides and serve as a tool for tracking microplastics origins (e.g., $\delta^{13}C$ links marine debris to specific polymer manufacturers). Besides, EA-IRMS is used in blue carbon sequestration and storage quantification as it allows for determining the total carbon, which is crucial for calculating organic and inorganic carbon. This is crucial for assessing the potential of climate change mitigation by coastal blue carbon ecosystems (seagrasses, mangroves, and salt marshes).

2.5 Conclusion

Nuclear and isotopic methods emerge as important solutions, providing reliability, precision, and distinctive insights into pollutant dynamics and thus driving pollution mitigation by allowing pollutants source identification enabling taking accurate regulatory actions and measure.

Techniques such as NAA and XRF when used together with radiometric dating approach allows for history reconstruction of pollutants such as heavy metals and rare earth elements and thus identification of their potential sources which is crucial for management projects and mitigations measure implementation. In this sense, combining nuclear and isotopic approaches with traditional methods such as ICP-MS enable scientists and policymakers to gain a multidimensional understanding of pollution, driving targeted, cost-effective mitigation strategies.

Stable isotopes ($\delta^{13}C$, $\delta^{15}N$) enable critical application where they allow for tracing pollutants origins and distinction between agricultural runoff and industrial discharges. Radiocarbon dating distinguishes fossil-derived pollutants, helping in tracking oil spill source, while stable isotope analysis reveals microplastic polymer origins, advancing waste management actions and strategies.

Organic and inorganic pollutants in marine and coastal environments affect biodiversity and accelerate climate changes effects. Addressing these issues require innovative and developed approaches that trace sources, evaluate impacts and track changes over years which allow better implementation of mitigation strategies. Nuclear and isotopic techniques emerge as essential tools for advancing the understanding of pollutants deposition history and behavior enabling guiding monitoring programs, regulating pollutants, restoration efforts and safeguarding life below water in good accordance with sustainable development goal SDG16.

References

Aarkrog A (1998) A retrospect of anthropogenic radioactivity in the global marine environment. Radiat Prot Dosim 75(1–4):23–31

Aggarwal SK (2016) Alpha-particle spectrometry for the determination of alpha emitting isotopes in nuclear, environmental and biological samples: past, present and future. Anal Methods 8(27):5353–5371

Antsiferova AA, Buzulukov YP, Demin VA, Demin VF, Rogatkin DA, Petritskaya EN et al (2015) Radiotracer methods and neutron activation analysis for the investigation of nanoparticle biokinetics in living organisms. Nanotechnol Russia 10:100–108

Barbieri F, Zavattini B, Marchetto D, de Ferri L, Pojana G (2023) Surface sea water monitoring of micro- and mesoplastics in a northern Adriatic coastal area. In: 18th International conference on chemistry and the environment-book of abstracts. European Chemical Society, pp 506–506

Benkdad A, Laissaoui A, El Bari H, Benmansour M, IbnMajah M (2008) Partitioning of radiostrontium in marine aqueous suspensions: laboratory experiments and modeling studies. J Environ Radioact 99(4):748–756

Benkdad A, Laissaoui A, Tornero MV, Benmansour M, Chakir E, Garrido IM, Moreno JB (2011) Trace metals and radionuclides in macroalgae from Moroccan coastal waters. Environ Monit Assess 182:317–324

Berto D, Rampazzo F, Gion C, Noventa S, Formalewicz M, Ronchi F et al (2019) Elemental Analyzer/Isotope Ratio Mass Spectrometry (EA/IRMS) as a tool to characterize plastic polymers in a marine environment. IntechOpen. https://doi.org/10.5772/intechopen.81485

Bisinger T, Hippler S, Michel R, Wacker L, Synal HA (2010) Determination of plutonium from different sources in environmental samples using alpha-spectrometry and AMS. Nucl Instrum Methods Phys Res Sect B Beam Interact Mater Atoms 268(7–8):1269–1272

Blanchard RL, Leddicotte GW, Moeller DW (1959) Water analysis by neutron activation. J AWWA 51(8):967–980

Choppin G, Liljenzin JO, Rydberg J, Ekberg C (2013) Radiochemistry and nuclear chemistry. Academic

Conrad ME, Daley PF, Fischer ML, Buchanan BB, Leighton T, Kashgarian M (1997) Combined ^{14}C and $\delta^{13}C$ monitoring of in situ biodegradation of petroleum hydrocarbons. Environ Sci Technol 31(5):1463–1469

Das DD, Sharma N, Chawla PA (2024) Neutron activation analysis: an excellent nondestructive analytical technique for trace metal analysis. Crit Rev Anal Chem 54(7):2450–2466

De-la-Torre GE, Aragaw TA (2021) What we need to know about PPE associated with the COVID-19 pandemic in the marine environment. Mar Pollut Bull 163:111879. https://doi.org/10.1016/j.marpolbul.2020.111879

Eisenbud M, Gesell T (1997) Environmental radioactivity: from natural, industrial, and military sources.4th edn. Academic, London

El-Taher A, Abdelhalim MAK (2014) Elemental analysis of soils from Toshki by using instrumental neutron activation analysis techniques. J Radioanal Nucl Chem 300:431–435

Ferrand E, Eyrolle F, Radakovitch O, Provansal M, Dufour S, Vella C et al (2012) Historical levels of heavy metals and artificial radionuclides reconstructed from overbank sediment records in lower Rhône River (South-East France). Geochim Cosmochim Acta 82:163–182

Ghosh S, Sharma I, Nath S, Webster TJ (2020) Bioremediation—the natural solution. In: Microbial ecology of wastewater treatment plants, pp 11–40. https://doi.org/10.1016/B978-0-12-822503-5.00018-7

Greenberg RR, Kingston HM (1983) Trace element analysis of natural water samples by neutron activation analysis with chelating resin. Anal Chem 55(7):1160–1165

Gustafsson O, Krusa M, Zencak Z, Sheesley RJ, Granat L, Engstrom E et al (2009) Brown clouds over South Asia: biomass or fossil fuel combustion? Science 323(5913):495–498

IAEA (2024) A guide to the use of sediments in reconstructing the pollution history of coastal areas. IAEA-TECDOC-1953. IAEA, Vienna. https://www.iaea.org/publications/15513/a--guide-to-the-use-of-sediments-in-reconstructing-the-pollution-history-of-coastal-areas. Accessed 30 Apr 2025

IAEA (2025) Studies of temporal trends of pollution in selected coastal areas by the application of isotopic and nuclear techniques. IAEA-TECDOC-2070. IAEA, Vienna. https://doi.org/10.61092/iaea.q1tx-6r1g

Jia G, Jia J (2012) Determination of radium isotopes in environmental samples by gamma spectrometry, liquid scintillation counting and alpha spectrometry: a review of analytical methodology. J Environ Radioact 106:98–119

Kaufman WJ, Orlob GT (1956) Measuring ground water movement with radioactive and chemical tracers. J AWWA 48(5):559–572

Kendall C, Elliott EM, Wankel SD (2007) Tracing anthropogenic inputs of nitrogen to ecosystems. In: Stable isotopes in ecology and environmental science. Wiley, pp 375–449

Kershaw PJ, Baxter AJ (1993) The transfer of reprocessing wastes from N.W. Europe to the Arctic. Deep-Sea Res 42:1413–1448

Laissaoui A, Benmansour M, Ziad N, Ibn Majah M, Abril JM, Mulsow S (2008) Anthropogenic radionuclides in the water column and a sediment core from the Alboran Sea: application to radiometric dating and reconstruction of historical water column radionuclide concentrations. J Paleolimnol 40:823–833

Laissaoui A, Mas JL, Hurtado S, Ziad N, Villa M, Benmansour M (2013) Radionuclide activities and metal concentrations in sediments of the Sebou Estuary, NW Morocco, following a flooding event. Environ Monit Assess 185:5019–5029

Laissaoui A, Mejjad N, Ziad N et al (2018) Evidence for a recent increase in delivery of atmospheric ^{210}Pb to Oualidia lagoon, coastal Morocco. Environ Monit Assess 190:642. https://doi.org/10.1007/s10661-018-7046-z

Li A (2011) Determination of the geographical origin of rice using stable isotope ratio and multi-element analyses. Doctoral dissertation, University of York

Lichtfouse E, Budzinski H, Garrigues P, Eglinton TI (1997) Ancient polycyclic aromatic hydrocarbons in modern soils: ^{13}C, ^{14}C and biomarker evidence. Org Geochem 26(5–6):353–359

Ligero RA, Barrera M, Casas-Ruiz M, Sales D, López-Aguayo F (2002) Dating of marine sediments and time evolution of heavy metal concentrations in the Bay of Cádiz, Spain. Environ Pollut 118(1):97–108

Livingston HD (1988) The use of Cs and Sr isotopes as tracers in the Arctic and Mediterranean Seas. Phil Trans R Soc Lond A325:161–176

Livingston HD, Povinec PP (2000) Anthropogenic marine radioactivity. Ocean Coast Manag 43(8–9):689–712

Mancinelli G, Vizzini S (2015) Assessing anthropogenic pressures on coastal marine ecosystems using stable CNS isotopes: state of the art, knowledge gaps, and community-scale perspectives. Estuar Coast Shelf Sci 156:195–204. https://doi.org/10.1016/j.ecss.2014.11.030

Marchant HK, Mohr W, Kuypers MM (2016) Recent advances in marine N-cycle studies using 15N labeling methods. Curr Opin Biotechnol 41:53–59. https://doi.org/10.1016/j.copbio.2016.04.019

Marın A, Lopez-Gonzalvez A, Barbas C (2001) Development and validation of extraction methods for determination of zinc and arsenic speciation in soils using focused ultrasound: application to heavy metal study in mud and soils. Anal Chim Acta 442(2):305–318

Mejjad N, Laissaoui A, El-Hammoumi O et al (2016) Sediment geochronology and geochemical behavior of major and rare earth elements in the Oualidia Lagoon in the western Morocco. J Radioanal Nucl Chem 309:1133–1143. https://doi.org/10.1007/s10967-016-4714-8

Mejjad N, Laissaoui A, El-Hammoumi O et al (2018) Geochemical, radiometric, and environmental approaches for the assessment of the intensity and chronology of metal contamination in the sediment cores from Oualidia lagoon (Morocco). Environ Sci Pollut Res 25:22872–22888. https://doi.org/10.1007/s11356-018-2370-y

Mejjad N, Laissaoui A, Benmhammed A, Fekri A, El Hammoumi O, Benkdad A et al (2022) Potential ecological risk assessment of rare earth elements in sediments cores from the Oualidia lagoon, Morocco. Soil Sediment Contam 31(8):941–958

Mejjad N, Laissaoui A, Fekri A et al (2023) Marine plastic pollution in Morocco: state of the knowledge on origin, occurrence, fate, and management. Environ Sci Pollut Res 30:107371–107389. https://doi.org/10.1007/s11356-023-26973-8

Mejjad N, Safhi AEM, Laissaoui A (2025) Insightful analytical review of potential impacts of microplastic pollution on coastal and marine ecosystem services. J Hazard Mater Adv 17:100578
Mishra RK, Mentha SS, Misra Y, Dwivedi N (2023) Emerging pollutants of severe environmental concern in water and wastewater: a comprehensive review on current developments and future research. Water Energy Nexus 6:74–95. https://doi.org/10.1016/j.wen.2023.08.002
Mok WJ, Ghaffar MA, Noor MIM, Lananan F, Azra MN (2023) Understanding climate change and heavy metals in coastal areas: a macroanalysis assessment. Water 15(5):891
Norris M (2015) Reconstruction of historic fossil CO_2 emissions using radiocarbon measurements from tree rings. Doctoral dissertation, Open Access Te Herenga Waka-Victoria University of Wellington
Noureddine A, Benkrid M, Maoui R, Menacer M, Boudjenoun R, Kadi-Hanifi M et al (2008) Radionuclide tracing of water masses and processes in the water column and sediment in the Algerian Basin. J Environ Radioact 99(8):1224–1232
Ohta A, Siraishi F, Nagahara T, Tomura K (1996) Neutron activation analysis of biological samples. Nihon Rinsho 54(1):207–214
Oladebeye AO (2017). Assessment of heavy metals in Nigerian vegetables and soils in Owo and Edo axes using X-ray fluorescence (XRF) technique. B.Sc. Project, Achievers University, Owo, Ondo State
Omokheyeke O, Sikoki FD, Laissaoui A, Akpuluma D, Onyagbodor PO, Benkdad A, Benmansour M (2014) Sediment geochronology and spatio-temporal and vertical distributions of radionuclides in the Upper Bonny Estuary (South Nigeria). Geochronometria 41:369–376
Ory C, Leboulleux S, Salvatore D, Le Guen B, De Vathaire F, Chevillard S, Schlumberger M (2021) Consequences of atmospheric contamination by radioiodine: the Chernobyl and Fukushima accidents. Endocrine 71:298–309
Pan Y (2023) Hazards of marine pollution to marine organisms and measures to mitigate the hazards. Highlights Sci Eng Technol 69:280–285. https://doi.org/10.54097/hset.v69i.11915
Pandiyan J, Mahboob S, Govindarajan M, Al-Ghanim KA, Ahmed Z, Al-Mulhm N et al (2021) An assessment of level of heavy metals pollution in the water, sediment and aquatic organisms: a perspective of tackling environmental threats for food security. Saudi J Biol Sci 28(2):1218–1225
Perini M, Bontempo L (2022) Liquid chromatography coupled to isotope ratio mass spectrometry (LC-IRMS): a review. TrAC Trends Anal Chem 147:116515. https://doi.org/10.1016/j.trac.2021.116515
Risal SY, Islam S, Ferdous J, Siddik MNA, Bakshi PK (2024) Natural radioactivity and heavy metal contamination in edible fish, shellfish and mollusks at the Bay of Bengal, Kuakata, Bangladesh. Heliyon 10(18)
Sable H, Singh V, Kumar V, Roy A, Pandit S, Kaur K et al (2024) Toxicological and bioremediation profiling of nonessential heavy metals (mercury, chromium, cadmium, aluminium) and their impact on human health: a review. Toxicol Anal Clin 36:205–234
Sandifer PA, Sutton-Grier AE (2014) Connecting stressors, ocean ecosystem services, and human health. Nat Resour Forum 38(3):157–167
Santhanabharathi B, Pradhoshini KP, Suhail Ahmed M, Priyadharshini M, Shafeeka Parveen MH, Alam L et al (2025) Source, fate and transfer of primordial radionuclides as potential contaminants in environmental matrices of high and low background radiation areas—a critical review. Int J Environ Anal Chem 105(4):954–980
Schwarzbauer J, Jovančićević B (2024) Isotopes in organic geochemistry. Springer International Publishing
Scrimgeour CM, Robinson D (2003) Stable isotope analysis and applications. In: Soil and environmental analysis. CRC Press, pp 348–393
Shah SB (2021) Heavy metals in the marine environment—an overview. In: Heavy metals in Scleractinian corals. Springer, pp 1–26

Shahjahan M, Taslima K, Rahman MS, Al-Emran MD, Alam SI, Faggio C (2022) Effects of heavy metals on fish physiology—a review. Chemosphere 300:134519

Soto-Navarro J, Jordá G, Compa M, Alomar C, Fossi MC, Deudero S (2021) Impact of the marine litter pollution on the Mediterranean biodiversity: a risk assessment study with focus on the marine protected areas. Mar Pollut Bull 165:112169. https://doi.org/10.1016/j.marpolbul.2021.112169

Stankovic S, Kalaba P, Stankovic AR (2014) Biota as toxic metal indicators. Environ Chem Lett 12:63–84

Stoeckl N, Adams V, Baird R, Boothroyd A, Costanza R, Hatton MacDonald D et al (2024) The value of Antarctic and Southern Ocean ecosystem services. Nat Rev Earth Environ 5(3):153–155

Swee-Li Yee M, Hii W, Looi CK, Lim M, Wong F, Kok Y, Tan K, Wong Y, Leong O (2021) Impact of microplastics and nanoplastics on human health. Nanomaterials 11(2):496. https://doi.org/10.3390/nano11020496

Szidat S, Jenk TM, Gäggeler HW, Synal HA, Hajdas I, Bonani G, Saurer M (2004) THEODORE, a two-step heating system for the EC/OC determination of radiocarbon (^{14}C) in the environment. Nucl Instrum Methods Phys Res Sect B Beam Interact Mater Atoms 223:829–836

Tan K, Cai X, Tan K, Kwan KY (2023) A review of natural and anthropogenic radionuclide pollution in marine bivalves. Sci Total Environ 896:165030

Tanizaki Y (1990) Neutron activation analysis of water samples. Activat Anal 2:377–405

Tavengwa NT, Moyo B, Musarurwa H, Dalu T (2021) Challenges and future directions in the analysis of emerging pollutants in aqueous environments. In: Emerging freshwater pollutants, pp 373–379. https://doi.org/10.1016/B978-0-12-822850-0.00016-8

Thiagarajan C, Devarajan Y (2024) The urgent challenge of ocean pollution: impacts on marine biodiversity and human health. Reg Stud Mar Sci 2024:103995

Tsabaris C, Kousidou G, Androulakaki EG, Patiris DL, Pappa FK, Marmara D, Krasakopoulou E (2024) Reconstruction of radioactivity and microplastic particles studying sedimentation/accumulation in the deep Lemnos basin, North Aegean Sea, Greece. J Soils Sediments 24(12):3938–3952

Walker TW, Janssens IA, Weedon JT, Sigurdsson BD, Richter A, Peñuelas J, Leblans NI, Bahn M, Bartrons M, De Jonge C, Fuchslueger L, Gunnarsdóttir GE, Oddsdóttir ES, Ostonen I, Poeplau C, Prommer J, Radujković D, Sardans J, Sigurðsson P et al (2019) A systemic overreaction to years versus decades of warming in a subarctic grassland ecosystem. Nat Ecol Evol 4(1):101–108. https://doi.org/10.1038/s41559-019-1055-3

Wan X, Lei M, Chen T (2016) Cost–benefit calculation of phytoremediation technology for heavy-metal-contaminated soil. Sci Total Environ 563:796–802

Wang L, Jin Y, Weiss DJ, Schleicher NJ, Wilcke W, Wu L, Guo Q, Chen J, O'Connor D, Hou D (2021) Possible application of stable isotope compositions for the identification of metal sources in soil. J Hazard Mater 407:124812. https://doi.org/10.1016/j.jhazmat.2020.124812

Wang Z, Luo P, Zha X, Xu C, Kang S, Zhou M et al (2022) Overview assessment of risk evaluation and treatment technologies for heavy metal pollution of water and soil. J Clean Prod 379:134043

West M, Ellis AT, Potts PJ, Streli C, Vanhoof C, Wobrauschek P (2015) 2015 Atomic Spectrometry Update—a review of advances in X-ray fluorescence spectrometry and their applications. J Anal At Spectrom 30(9):1839–1889

Witkowska E, Szczepaniak K, Biziuk M (2005) Some applications of neutron activation analysis: a review. J Radioanal Nucl Chem 265:141–150

Zaborska A, Carroll J, Papucci C, Pempkowiak J (2007) Intercomparison of alpha and gamma spectrometry techniques used in ^{210}Pb geochronology. J Environ Radioact 93(1):38–50

Zaim N, Dogan C, Camtakan Z (2016) Neutron activation analysis of soil samples from different parts of Edirne in Turkey. J Appl Spectrosc 83:271–276

Zhang G, Liu J, Li J, Li P, Wei N, Xu B (2021) Radiocarbon isotope technique as a powerful tool in tracking anthropogenic emissions of carbonaceous air pollutants and greenhouse gases: a review. Fundam Res 1(3):306–316. https://doi.org/10.1016/j.fmre.2021.03.007

Chapter 3
Application of Radiotracer Techniques to Trace Pollutants in Marine Ecosystems

Hajar Khabouche, Jamal Mabrouki, Driss Azdem, Selma Mabrouki, Salah Eddine Sbai, Miloudia Slaoui, Imane Rouichat, and Houria Al Bakraoui

Abstract Numerous contemporary environmental concerns jeopardizing the health and sustainability of coastal and marine ecosystems are induced or intensified by a multitude of anthropogenic stressors, further aggravated by climate and oceanic changes. Since the public became aware of the growing contamination of water supplies, deterioration of the atmosphere, and poisoning of soils by industrial toxicants, environmental pollution issues have grown to be global issues. The use of biomonitoring techniques in assessing and controlling strategies for the marine environment has several advantages (precise, responsive, and feasible) over any other technique method. Nuclear techniques have opened up new perspectives on the pathways and rates of uptake. This chapter focuses on clarifying the function of radiological marine biomonitoring. This paper highlights the application of radiotracers, illustrates the advancement of existing radioecological tracing techniques for contemporary environmental challenges, and presents insights on forthcoming difficulties that could benefit from both current and novel radiotracer methodologies.

Keywords Radiotracers · Radioecology · Ecosystem condition · Marine · Water contamination

H. Khabouche · M. Slaoui · I. Rouichat · H. Al Bakraoui
Energy, Materials and Sustainable Development (EMDD) Laboratory—Higher School of Technology Sale, Center of Water, Natural Resources, Environment and Sustainable Development (CERN2D), University Mohammed V, Rabat, Morocco

J. Mabrouki (✉) · D. Azdem · S. E. Sbai
Laboratory of Spectroscopy, Molecular Modelling, Materials, Nanomaterials, Water and Environment, CERN2D, Faculty of Science, Mohammed V University, Rabat, Morocco
e-mail: jamal.mabrouki@um5r.ac.ma

S. Mabrouki
Laboratory of Marine Biotechnology and Environment—CNRST Labeled Research Unit, Faculty of Sciences, Chouaib Doukkali University, El Jadida, Morocco

N. Mejjad (ed.), *Nuclear-Based Approaches application for marine pollution assessment and monitoring*, Environmental Science and Engineering,
https://doi.org/10.1007/978-3-032-12197-4_3

3.1 Introduction

The negative effect on marine ecosystems is believed that it is caused by pollution that originate naturally as well as through humans. This incident emphasizes the necessity of having an in-depth understanding of dispersion, bioaccumulation and the subsequent consequences of these contaminants on marine mammals. Radiotracer methods have been a key method of studying chemical processes, and they have had minimal effect on the ecosystems being studied. They are consequently regarded as vital tools that could be employed to safeguard these ecosystems from the effects of water pollution (NEW CRP 2018; Fesenko et al. 2020). Due to the availability of radionuclides in the soil and water, the biological processes can be subjected to significant disturbance. The contamination may be caused by a host of factors including the natural process, the testing of nuclear weapons, care management as well as disasters like the Chernobyl 1986 incident and the 2011 Fukushima accident that highlighted the difficulty of bombs of nuclear fuel. These properties of these contaminants pose a risk, due to their chemical characteristics (they are highly soluble, readily transferable, long half-lives) and their ability to be absorbed by living organisms (Burger and Lichtscheidl 2019; Thakur and Kumar 2024). Radioactive isotopes are commonly referred to as radiotracers. They play a part and serve the function of tracing the pathways and effects of pollution within the ecosystem. They have distinctive characteristics. Isotopes may be utilized by scientists in their studies of issues as biomagnification of components in the nutrition cycle of food webs and also the spread of pollutant. They are suitable for studying marine habitats in actual field conditions, mesocosms in controlled conditions, as well as in laboratory as well as environment. Their work concentrates on radiation absorption by radionuclides (strontium as well as caesium) as well as the heavy metals (mercury and Cadmium) within marine organisms as well as sediments (Shen et al. 2012; Kim et al. 2021). It's been demonstrated that when used correctly radiotracers don't pose human health risks and permit researchers to research the causes and impacts of environmental pollution more deeply (Ammar et al. 2024). Incorporating the studies of radiotracers together with analytical methods including liquid chromatography and mass spectrometry, has led in our knowledge of how marine life is affected by pollution. Radiotracers can help us understand the processes of flux of nutrients that are involved in the acidification of oceans and the production of algal blooms that pose danger to humans (Tan et al. 2023). But, the use of radiotracers needs careful considerations which will ensure their safety to the natural environment in addition to reducing risk to the environment. This chapter explores their application in environmental studies in particular, the most recent advances in techniques for radioecological tracer which have been developed in response to the present environmental challenges.

3.2 Radiotracers and Marine Ecosystems

3.2.1 Marine Ecosystems

The marine ecosystem is a vital component of the Earth's ecosystem because it provides a variety of products and services to people, supports the economic growth of human societies and improves people's quality of life (Wang et al. 2024). Terrestrial ecosystems are generally easier to monitor, however because of their high level of dynamism and their 3D environment marine ecosystems suffer from the absence of spatial data. In this sense the concept of mapping is recognized as an essential tool for the application for ecosystem services (ES) and modeling is a key instrument for assessing the value of ecosystem services in marine ecosystems (Ilieva et al. 2022; Addamo et al. 2024). Marine ecosystems are intricate systems made up of interdependent biological, chemical, and physical elements. Introduced pollutants have a variety of biological and environmental effects as a result of their dynamic interactions with water, sediments, and organisms. These interactions have been clarified thanks in large part to radiotracer techniques, which offer in-depth understanding of pollutant behavior (Chalkiadakis et al. 2022; Dasgupta 2024). A marine ecology is depicted in Fig. 3.1.

Chemical contaminants resulting from human activity both on land and in the sea may enter the marine environments. It can be a result of the diffuse source (widespread pollution) like streams, run-off and long-range atmospheric deposition and also from point pollution sources (localized pollution) like discharges or accidental

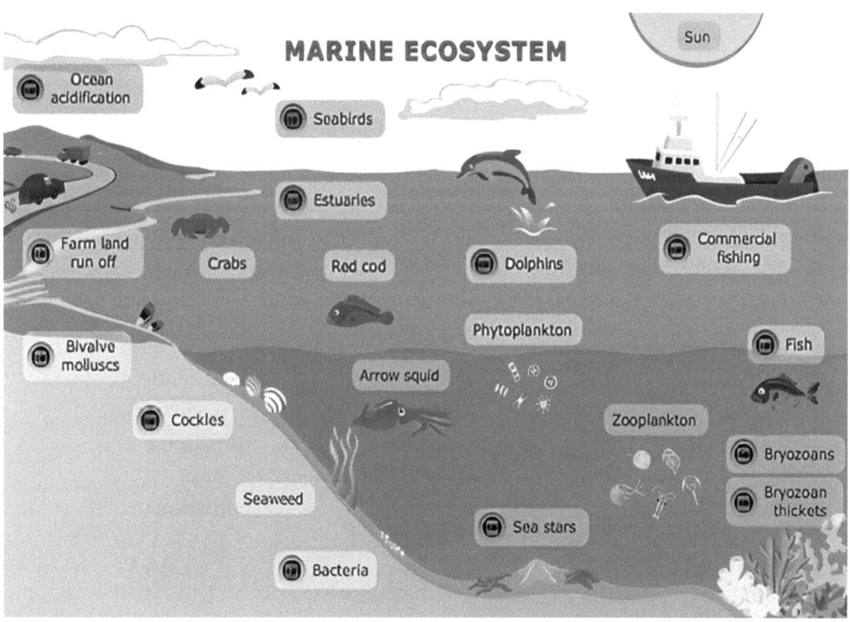

Fig. 3.1 Various parts of the marine ecosystem

spills. Examples of contaminants include nutrients that may encourage the growth of undesirable algae and a range of dangerous substances, many of which remain toxic to living organisms or bioaccumulative. The compounds of greatest concern are identified through national and international prioritization and their levels are monitored in sediment, water and biota. Trace metals, hydrocarbons, pesticides, dioxins, PCBs, flame retardants and endocrine disrupting compounds are some examples of substances monitored by the Monitoring Initiative.

3.2.2 *Radiotracers*

Radiotracers are substances used as markers to determine the location of a compound or to track its progression through a physical process or chemical reaction. Ideally, tracers exhibit similar chemical properties to the target substance, although their physical properties may differ depending on the system under study (Xia et al. 2020; Kusch et al. 2021). A tracer is a material to track or trace a physiological or biological process. Tracers may be radionuclides or can be, more often, small molecules or larger biomolecules (such as peptides and antibodies) containing radionuclides. Such labeled compounds are also called radiotracers or radiopharmaceuticals (Fig. 3.2). A radiotracer is a chemical compound whereby one or more atoms have been substituted with a radioisotope. It is possible to use radiotracers to track radioactivity decay and gain good understanding of chemical reactions mechanisms. The latter are also popular in the visualization of flow studies using such technologies as Computer Automated Radioactive Particle Tracking (CARPT), Positron Emission

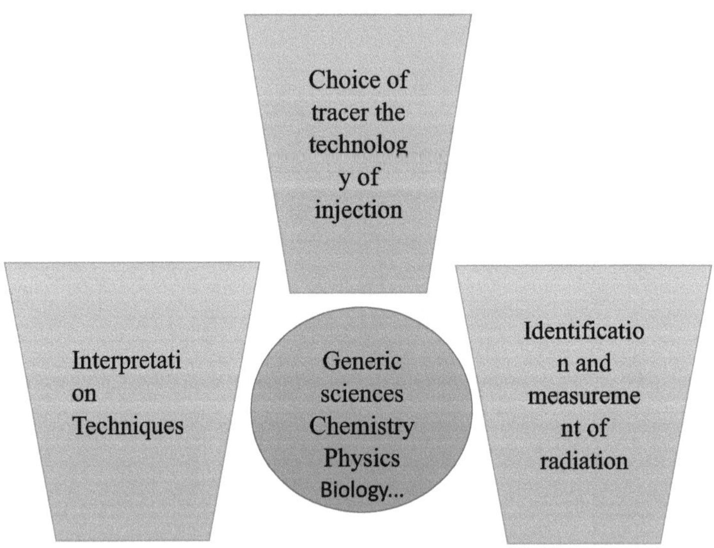

Fig. 3.2 Radiotracer technology and related elements

Tomography (PET), and Single-Photon Emission Computed Tomography (SPECT). Radioactive tracers have been widely applied to study axonal transport, and especially fast axonal transport. Radiotracers are well suited to the study of the movement and transformation of substances in complex ecosystems because of their distinctive properties (radioactive decay patterns, isotopic signatures, and detectable alpha or beta emissions) and the fact that they can be specifically introduced into the habitat and traced through space and time (Österblom et al. 2015; Owuor et al. 2017).

Since radiotracers have a half-life, the time it takes half of the isotopes to decay to a stable form the process of their isotope decay is predictable. Moreover, a gamma spectrometer, or scintillation counter, can be used to detect their radioactive emissions at high sensitivity (Andersen et al. 2004; Bali and Sidhu 2021; Bahsaine et al. 2024). The mechanistic characteristics of the radiotracers enable them to take advantage of their chemical and physical characteristics to communicate with the environment. Specifically, the isotopes such as the ^{14}C can be applied to study the carbon cycle and ^{65}Zn is used to analyze the bioavailability of the heavy metals because they are found in organic compounds. These properties allow radiotracers offer important insight into the environmental health and the function of ecosystems through revealing the smallest details of processes like the migration, sedimentation, and bioaccumulation of pollutants (Hédouin et al. 2016; Cabrerizo et al. 2022). The development and research of technology is ongoing and ongoing, even though radioisotopes have been used to resolve industrial issues for over 50 years. This constant attention to radioisotopes can be explained by two major reasons. Because radioactive isotopes offer insight that isn't possible using other methods the radiotracer technology is influenced principally by the requirements of the industrial. These isotopes allow real-time process monitoring without interrupting operations, which has significant economic benefits, including avoiding plant shutdowns and optimizing production. In addition, the development of radiotracer methodology involves a wide range of scientific and technological fields, such as radioisotope production, radiation detection, data collection and analysis, and mathematical modelling. Improvements in any of these areas lead to further advances in radiotracer applications, increasing their effectiveness and broadening their potential applications across a range of industries.

3.3 Trace Element Effects and Influence of Tracking Radioisotopes

In both terrestrial and aquatic natural systems, the transit, mobility and segregation of metallic and metalloid trace elements depend on the chemical composition of the element, which is ultimately influenced by the physico-chemical and biological characteristics of the system. Trace elements can bind to clay minerals, organic molecules, hydrated Fe and Mn oxides and other materials in a variety of ways once they reach the sediment. We now know more about trace elements thanks to the development of specialized analytical methods, field research, controlled laboratory tests and geochemical modelling (Baran et al. 2019; Gao et al. 2018). Exchange, sorption, complexation, or chelation are the processes that allow organic matter (OM) to incorporate trace metals. The characteristics and level of alteration of trace elements determine the potential of OM in examining their mobility. Measurement of trace element (such as Mn Fe as well as Sr Zr, Rb, and Ti) and details about the nature and sources of OM are possible due to the development in XRF technology. Locating potential OM sources is also becoming more vital due to the cation exchange capabilities of OM and functions as agent chelating for trace elements (Cheng et al. 2021). The geochemical tracing field has seen a significant increase in popularity following the discovery of large quantities of ^{90}Sr, ^{137}Cs, ^{3}H and ^{14}C released into the oceans as part of nuclear weapon testing around the middle of 1950s. The shift in chemical oceanographers led them to favour radiotracers over conventional chemical tracer. Contrary to traditional tracers' radiotracers offered unique insights into oceanic processes, which allowed scientists to obtain crucial data without relying on speculations about the bioactivity rates that occur in marine environments. The patterns of distribution of these isotopes proved to be an important tool for understanding the behaviour of ocean systems. This includes the movement of water masses mixing rates and biogeochemical cycles (Camusso et al. 1999; Cartwright and Morgenstern 2018).

The radionuclides most commonly associated with nuclear accidents are ^{90}Sr and ^{137}Cs. As a fission by-product of uranium and plutonium, Sr diffuses more widely through the atmosphere than other nuclides. These two radionuclides have adverse effects on human health. Radioactive Sr is harmful to humans because of its physical similarity to calcium (Ca). Because of this similarity, leukemia and bone marrow cancer can result from the accumulation of Sr in the bone, just as Ca does (Tian et al. 2023; Bahsaine et al. 2024). Cesium (Cs), a long-lived fission product with a half-life of approximately 30 years, represents a major environmental and health concern. Its chemical similarity to potassium facilitates rapid absorption and accumulation in soft tissues and muscles, leading to prolonged internal irradiation. Chronic exposure to Cs is associated with an elevated risk of cancers, cardiovascular disorders, and immune system impairment. Its persistence in ecosystems underscores its dual role as both a valuable tracer for environmental studies and a marker of long-term anthropogenic contamination with significant ecological and public health implications. In aquatic systems, radionuclides can be used to analyze the kinetics of physical, chemical, biological, or geological processes. Due to fallout from nuclear weapon tests, a remarkable amount of radioactive material has been deposited on the earth's surface (Jung et al. 2020 ; Khan et al. 2020). The isotopes that are radioactively enriched may be released to the surrounding environment and domesticated to determine the biogeochemical cycling of different elements. Tritium (^{3}H), carbon 14 (^{14}C), and chlorine 36 (^{36}Cl) have been the most common radioactive isotopes used since a long time. More recently, as more complex analytical methods were developed, radioactive isotopes of noble gases, including argon 39 (^{39}Ar), krypton 81 (^{81}Kr) and krypton 85 (^{85}Kr), have become useful tracers to measure water and atmospheric residence times.

3.4 Marine Radiological Biomonitoring

Radiological marine biomonitoring is of great assistance with the control and evaluation of radioactive pollution in marine ecosystems. The radionuclides may be observed and quantified on the marine animals, the sediments and water in order to ascertain the origin of contamination, such as the nuclear tests or industrial effluents. By analysing the action of trophic accumulation and bio accumulation. This is a method of preserving the ecology of the environment. Radiotracer techniques are also used to assess the impact of radionuclides in the ecosystem of marine life as well as the food web and their organization. In addition, they aid in ensuring compliance to the rules, providing vital information on standard of environmental quality and safeguarding public health by providing support of an assessment of the safety of seafood. Biomonitoring is one of the most important aspects of studies on the ecology of the ocean, since it's an important aspect for sustainable management of the ocean's resource. It helps in observing the processes of oceanography as well as to investigate the ecological effects of radioactive pollutants (Fig. 3.3) (Roveta et al. 2021; Al-Sharif et al. 2024; Cobbing 2013). It's a technique that involves the consistent building-up and examination of biochemical samples like tissues and fluids and analyzes the exposure of chemicals that are made by humans and natural sources.

Fig. 3.3 Illustration of marine radiological biomonitoring

Biomonitoring of changes to the environment as well as the ecological consequences of those changes provide an appropriate foundation from that it is possible to assess the change taking place in the world, through the examination of biochemical and physiological reactions of the organisms which serve as a signpost of the overall health of the ecological system (Osman and Kloas 2010). The multidisciplinary approach is widely applied to both the evaluation of environmental quality and the analysis of the effects of pollutants to which its definitions and scope may vary according to the purpose of the study. Initially, biomonitoring primarily was intended to determine the amount of heavy metals in a specific target area of the environment (Radiotracers 2016; Science Learning Hub 2024). Slowly but surely it has been applied to a broad range of approaches to the study of the deposition of pollutants, ecological impacts and the long-term environmental change across a scale of ecological magnitude (Zhu et al. 2025). The physical, biological, climate, and physico-chemical indicators (pH, trace metals, and radionuclides) used as examples of biomonitoring allow obtaining a full picture of the dynamics of pollutants. Unlike pure physical or chemical tests, it offers essential clues to the bioavailability of pollutants and biotransformation process, and to the effects of low-dose exposures. It is a better way to measure the impact of pollution because it is more sensitive and economical, and it captures the cumulative and interaction effects on organisms and ecosystems over time periods (Al-Sharif et al. 2023a, b). Nuclear power, characterized as intense thermal energy source, is notable for being one of the lowest CO_2 emitting energy technologies, contributing about 13.5% of the world's electricity supply. Despite its environmental benefits in terms of emissions, nuclear energy poses significant risks to the biosphere, primarily through radiation emissions. These emissions can have profound effects on biological systems, as radiation damages critical molecular targets such as DNA. Hutchinson (1966) highlighted this concern, emphasizing that DNA is essential for cell replication and proliferation. Damage to DNA, if not repaired, can lead to mutations, genomic instability and cell death, underscoring the biological risks associated with nuclear power (Hutchinson 1966; Pioro and Duffey 2015).

3.5 Using Radiotracers to Detect Pollutants

The general principle of the radiotracer technique, as applied to continuous flow systems, is to inject a radioactive material, that mimics that of the process material, at the inlet of the system (Simoneit et al. 2017; Stewart et al. (2008); Tang et al. 2011). This tracer is then monitored at various points in the system using radiation detectors such as gamma counters or scintillation detectors. The tracking of tracer levels in strategically placed locations provides crucial information about system performance, which allows the identification of operational irregularities and the assessment of the hydrodynamic behavior of the process units. Similar methods are used in batch-mode systems, in which radiotracers are introduced along with its location, it is monitored across several locations to assess the flow dynamics, enhance process efficiency, and identify possible system failures (Pant 2022). Tracers that are used in industrial or research are generally classified as radiotracers or conventional tracers. Conventional tracers include chemicals such as electro-affinity activatable tracer and stable isotopic tracer, which are mostly used for lab-scale research in controlled, low-temperature environments. Although these tracers can be useful for small-scale research but their applications are restricted by their lower sensitivities and the possibility of system interference, making them ineffective for large-scale monitoring in industrial settings. However, radiotracers offer greater flexibility and accuracy that allow for real-time monitoring of processes that are dynamic within large systems, including oilfield operations and greater diagnostic capabilities (Kumar et al. 2017).

3.6 Advances in Radioecological Tracer Techniques

Currently the development of radioecological tracer methods is essential to resolve the pressing environmental problems, namely, the problems that concern nuclear accidents, radioactive pollution and the more extensive impact caused by the climate changes. Due to technological developments, there are currently highly sensitive and precise radiation detector instruments, including portable monitoring instruments, and advanced Gamma spectrometers that enable the identification of radionuclides in terribly low concentrations right in the setting.

The Inductively coupled plasma Mass Spectrometry (ICP-MS) and Accelerator Mass Spectrometry (AMS) are high-resolution and high-quality analytic systems that can provide believable methods of the determination of radionuclides in low concentrations with spectacular accuracy. Furthermore, the use of novel radiotracers such as Storuim (^{90}Sr) and technetium (Tc-99m) and a multi-isotope methodology can allow scientists to trace the pathways of contamination in complex systems and differentiate the naturally occurring and man-made sources of radioactivity unambiguously (Degger et al. 2011; Walsh et al. 2017; Balaram et al. 2023). Numerous sources of data on the environment can be merged using mathematical models. These could be utilized to better forecast the movements of radionuclides throughout the aquatic and terrestrial systems as well as the biological system. Such models could be enhanced using remote sensing technologies and, in conjunction in conjunction with Geographic Information Systems (GIS) can allow tracking patterns of pollution across the long period with large quantities (Marine Institute 2024; Marine Pollution 2024). The evolution of bioaccumulation research as well as precise estimations of human exposure enhance our understanding of how radionuclides are transported in food webs and the effects they have for human health and the environment particularly in post-nuclear catastrophe situations (Bergdahl and Skerfving 2022). It is clear that radioecological tracing can be useful to identify hotspots for pollution as well as for planning remediation as well as improving monitoring of environmental conditions in regions at high risk. In addition, it is becoming more likely to have the likelihood of monitoring radionuclides real time, due to international collaborations, made possible by WHO, IAEA, and WHO as well as assistance from open-access databases as well as projects by citizen scientists. Taken together they can be useful to improve the handling of radiations from the environment and gaining an understanding of the longer-term impacts of radionuclides to the environment and well-being of the people (Kavanagh et al. 2017; Braeuer et al. 2020; Kumar et al. 2023).

3.7 Conclusion

A reliable tracer ought to be abundant and chemically stable, and possessing the capability to give exact information regarding the particular geographic sources as well as biological source. Carbon (^{13}C) and nitrogen (^{15}N) as well as the ratio of nitrogen and carbon are proven as reliable indicators which have become an accepted method of environmental and ecological research. However, they can be influenced by numerous environmental variables, including the type of soil, temperature in vegetation areas, type of soil and the pre-established geochemical background. In addition to isotopic methods biomolecules, pigments and lipids and the structural components of cell membranes are being identified as long-lasting biomarkers. The development of radiotracers has increased our ability to monitor the routes of pollution and to trace how they change, as well as to predict how they affect the environment, in this case, aquatic and marine. Such tools do not only augment our knowledge about the dynamics of the contaminants, but also enhance monitoring strategies, that will eventually be beneficial in the effective management of the ecosystems in sustainable means in order to safeguard the ecosystem. Radiotracers can be used to detect low-concentration contaminants. These will also give important information on how the contaminants impact on marine organisms and overall health of marine ecosystems. To find out the cause of the pollution and to measure the efficiency of the mitigation mechanisms these procedures help researchers to trace the pathways along the way pollutants are transferred, gathered and, possibly, restored to the environment. Radiotracers are the important part of the research and they are important to evolve the specific actions to decrease pollution and to develop the evidence-based strategies to secure and preserve the marine biodiversity.

References

Addamo AM, La Notte A, Guillen J (2024) Status of mapping, assessment and valuation of marine ecosystem services in the European seas. Ecosyst Serv 67:101631. https://doi.org/10.1016/j.ecoser.2024.101631

Al-Sharif S, Chapara M, El-Taher A et al (2023a) Study of the synergistic effects of meteorological, oceanographic, biological, and radiological parameters (MOBR) on marine ecosystems around a nuclear power plant. Egypt J Aquat Biol Fish 27:1393–1413. https://doi.org/10.21608/ejabf.2023.324931

AL-Sharif SA, Chapara M, El-Taher A, Salah-Tantawy A (2023b) Utilizing fluctuation asymmetry to assess the effects of U-238 radioactivity on the marine ecosystem around the Madras nuclear power plant, India. Oceanol Hydrobiol Stud 52:437–445. https://doi.org/10.26881/oahs-2023.4.05

Al-Sharif S, Chapara M, El-Taher A, Osman A (2024) Biomonitoring of marine radioactive pollution: a review. Int J Nucl Energy Sci Technol 17:116–134. https://doi.org/10.1504/IJNEST.2024.10067189

Ammar A, Nouira A, El Mouridi Z, Boughribil S (2024) Recent trends in the phytoremediation of radionuclide contamination of soil by cesium and strontium: sources, mechanisms and methods: a comprehensive review. Chemosphere 359:142273. https://doi.org/10.1016/j.chemosphere.2024.142273

Andersen PK, Ghassemi A, Ghassemi M (2004) Nuclear waste. In: Cleveland CJ (ed) Encyclopedia of energy. Elsevier, New York, pp 449–463

Bahsaine K, Benzeid H, El Allaoui B et al (2024) Porous polyvinyl fluoride coated cellulose beads for efficient removal of Cd(II) from phosphoric acid. Int J Biol Macromol 254:127867. https://doi.org/10.1016/j.ijbiomac.2023.127867

Balaram V, Copia L, Kumar US et al (2023) Pollution of water resources and application of ICP-MS techniques for monitoring and management—a comprehensive review. Geosyst Geoenviron 2:100210. https://doi.org/10.1016/j.geogeo.2023.100210

Bali AS, Sidhu GPS (2021) Arsenic acquisition, toxicity and tolerance in plants—from physiology to remediation: a review. Chemosphere 283:131050. https://doi.org/10.1016/j.chemosphere.2021.131050

Baran A, Mierzwa-Hersztek M, Gondek K et al (2019) The influence of the quantity and quality of sediment organic matter on the potential mobility and toxicity of trace elements in bottom sediment. Environ Geochem Health 41:2893–2910. https://doi.org/10.1007/s10653-019-00359-7

Bergdahl IA, Skerfving S (2022) Chapter 19: Lead. In: Nordberg GF, Costa M (eds) Handbook on the toxicology of metals5th edn. Academic, pp 427–493

Braeuer S, Borovička J, Kameník J et al (2020) Is arsenic responsible for the toxicity of the hyperaccumulating mushroom Sarcosphaera coronaria? Sci Total Environ 736:139524. https://doi.org/10.1016/j.scitotenv.2020.139524

Burger A, Lichtscheidl I (2019) Strontium in the environment: review about reactions of plants towards stable and radioactive strontium isotopes. Sci Total Environ 653:1458–1512. https://doi.org/10.1016/j.scitotenv.2018.10.312

Cabrerizo MJ, Medina-Sánchez JM, González-Olalla JM et al (2022) Microbial plankton responses to multiple environmental drivers in marine ecosystems with different phosphorus limitation degrees. Sci Total Environ 816:151491. https://doi.org/10.1016/j.scitotenv.2021.151491

Camusso M, Balestrini R, Martinotti W, Arpini M (1999) Spatial variations in trace metal and stable isotope content of autochthonous organisms and sediments in the river Po system (Italy). Aquat Ecosyst Health Manage 2:39–53. https://doi.org/10.1080/14634989908656938

Cartwright I, Morgenstern U (2018) Using tritium and other geochemical tracers to address the "old water paradox" in headwater catchments. J Hydrol 563:13–21. https://doi.org/10.1016/j.jhydrol.2018.05.060

Chalkiadakis C, Drakou EG, Kraak M-J (2022) Ecosystem service flows: a systematic literature review of marine systems. Ecosyst Serv 54:101412. https://doi.org/10.1016/j.ecoser.2022.101412

Cheng Z, Yu F, Ruan X et al (2021) GDGTs as indicators for organic-matter sources in a small subtropical river-estuary system. Org Geochem 153:104180. https://doi.org/10.1016/j.orggeochem.2021.104180

Cobbing N (2013) Radiotracers: essential nuclear tools to understand oceans. IAEA Bulletin

Dasgupta P (2024) The economics of biodiversity. Cambridge University Press

Degger N, Wepener V, Richardson BJ, Wu RSS (2011) Application of artificial mussels (AMs) under South African marine conditions: a validation study. Mar Pollut Bull 63:108–118. https://doi.org/10.1016/j.marpolbul.2011.04.040

Fesenko S, Shinano T, Onda Y, Dercon G (2020) Dynamics of radionuclide activity concentrations in weed leaves, crops and of air dose rate after the Fukushima Daiichi Nuclear Power Plant accident. J Environ Radioact 222:106347. https://doi.org/10.1016/j.jenvrad.2020.106347

Gao L, Wang Z, Li S, Chen J (2018) Bioavailability and toxicity of trace metals (Cd, Cr, Cu, Ni, and Zn) in sediment cores from the Shima River, South China. Chemosphere 192:31–42. https://doi.org/10.1016/j.chemosphere.2017.10.110

Hédouin L, Metian M, Teyssié J-L et al (2016) Bioaccumulation of ^{63}Ni in the scleractinian coral Stylophora pistillata and isolated Symbiodinium using radiotracer techniques. Chemosphere 156:420–427. https://doi.org/10.1016/j.chemosphere.2016.04.097

Hutchinson F (1966) The molecular basis for radiation effects on cells. Cancer Res 26:2045–2052

Ilieva L, Ursano I, Traista L et al (2022) Biomimicry as a sustainable design methodology—introducing the 'biomimicry for sustainability' framework. Biomimetics 7:37. https://doi.org/10.3390/biomimetics7020037

Jung H et al (2020) Stable isotopes of water and nitrate for the identification of groundwater flow-paths: a review. Water 12(1):138

Kavanagh L, Keohane J, Cleary J et al (2017) Lithium in the natural waters of the south east of Ireland. Int J Environ Res Public Health 14:561. https://doi.org/10.3390/ijerph14060561

Khan H, Laas A, Marcé R et al (2020) Major effects of alkalinity on the relationship between metabolism and dissolved inorganic carbon dynamics in lakes. Ecosystems 3:1566–1580. https://doi.org/10.1007/s10021-020-00488-6

Kim H-S, Lee S-H, Jo HY et al (2021) Diversity and composition of soil Acidobacteria and Proteobacteria communities as a bacterial indicator of past land-use change from forest to farmland. Sci Total Environ 797:148944. https://doi.org/10.1016/j.scitotenv.2021.148944

Kumar R, Pant HJ, Goswami S et al (2017) Investigation of holdup and axial dispersion of liquid phase in a catalytic exchange column using radiotracer technique. Appl Radiat Isot 121:51–60. https://doi.org/10.1016/j.apradiso.2016.12.007

Kumar V, Lakkaboyana SK, Sharma N et al (2023) A critical assessment of technical advances in pharmaceutical removal from wastewater—a critical review. Case Stud Chem Environ Eng 8:100363. https://doi.org/10.1016/j.cscee.2023.100363

Kusch S, Wakeham SG, Sepúlveda J (2021) Diverse origins of "soil marker" bacteriohopane-polyols in marine oxygen deficient zones. Org Geochem 151:104150. https://doi.org/10.1016/j.orggeochem.2020.104150

Marine Institute (2024) Marine pollutants. Marine Institute. https://www.marine.ie/site-area/areas-activity/marine-environment/marine-pollutants. Accessed 2 Dec 2024

Marine Pollution (2024). https://education.nationalgeographic.org/resource/marine-pollution. Accessed 2 Dec 2024

NEW CRP (2018) Applied radioecological tracers to assess coastal and marine ecosystem health (K41019). https://www.iaea.org/newscenter/news/new-crp-applied-radioecological-tracers-to-assess-coastal-and-marine-ecosystem-health-k41019. Accessed 2 Dec 2024

Osman AGM, Kloas W (2010) Water quality and heavy metal monitoring in water, sediments, and tissues of the African catfish Clarias gariepinus (Burchell, 1822) from the river Nile, Egypt. J Environ Prot 1:389–400. https://doi.org/10.4236/jep.2010.14045

Österblom H, Jouffray J-B, Folke C et al (2015) Transnational corporations as 'keystone actors' in marine ecosystems. PLoS One 10:e0127533. https://doi.org/10.1371/journal.pone.0127533

Owuor MA et al (2017) Mapping of ecosystem services flow in Mida Creek, Kenya. Ocean Coast Manage 140:11–21. https://doi.org/10.1016/j.ocecoaman.2017.02.013

Pant HJ (2022) Applications of the radiotracers in the industry: a review. Appl Radiat Isot 182:110076. https://doi.org/10.1016/j.apradiso.2021.110076

Pioro I, Duffey R (2015) Nuclear power as a basis for future electricity generation. J Nucl Eng Radiat Sci 1:011001. https://doi.org/10.1115/1.4029420

Radiotracers (2016). https://www.iaea.org/topics/radiotracers. Accessed 2 Dec 2024

Roveta C, Annibaldi A, Afghan A et al (2021) Biomonitoring of heavy metals: the unexplored role of marine sessile taxa. Appl Sci 11:580. https://doi.org/10.3390/app11020580

Science Learning Hub (2024) Marine ecosystem. Science Learning Hub. https://www.science-learn.org.nz/image_maps/32-marine-ecosystem. Accessed 2 Dec 2024

Shen Y, Fichot CG, Benner R (2012) Floodplain influence on dissolved organic matter composition and export from the Mississippi—Atchafalaya River system to the Gulf of Mexico. Limnol Oceanogr 57:1149–1160. https://doi.org/10.4319/lo.2012.57.4.1149

Simoneit BRT, Pisani O, Ekpo BO et al (2017) Lipid biomarker analysis of suspended particulate matter from the Great Kwa River, SE Nigeria: origins and environmental implications of biogenic and anthropogenic organic compounds. Aquat Geochem 23:89–108. https://doi.org/10.1007/s10498-017-9311-0

Stewart GM et al (2008) Chapter 8: The bioaccumulation of U- and Th-series radionuclides in marine organisms. In: Radioactivity in the environment. Elsevier, pp 269–305

Tan K, Cai X, Tan K, Kwan KY (2023) A review of natural and anthropogenic radionuclide pollution in marine bivalves. Sci Total Environ 896:165030. https://doi.org/10.1016/j.scitotenv.2023.165030

Tang S, Liao S, Guo J, Song Z, Wang R, Zhou X (2011) Growth and cesium uptake responses of Phytolacca americana Linn. and Amaranthus cruentus L. grown on cesium contaminated soil to elevated CO_2 or inoculation with a plant growth promoting rhizobacterium Burkholderia sp. D54, or in combination. J Hazard Mater 198(198):188–197

Thakur A, Kumar A (2024) Emerging paradigms into bioremediation approaches for nuclear contaminant removal: from challenge to solution. Chemosphere 352:141369. https://doi.org/10.1016/j.chemosphere.2024.141369

Tian Y, Ma L, Gao N et al (2023) Removal of surface cesium ion contamination by peelable composite coating with a highly effective magnetic adsorbent. J Alloys Compd 958:170280. https://doi.org/10.1016/j.jallcom.2023.170280

Walsh JJ, Lenes JM, Weisberg RH et al (2017) More surprises in the global greenhouse: human health impacts from recent toxic marine aerosol formations, due to centennial alterations of world-wide coastal food webs. Mar Pollut Bull 116:9–40. https://doi.org/10.1016/j.marpolbul.2016.12.053

Wang J, Peng B, Huang L, Chen K (2024) Tracing offshore marine ecosystem asset changes based on physical accounting: a case of Xiamen Sea area. Ocean Coast Manag 259:107420. https://doi.org/10.1016/j.ocecoaman.2024.107420

Xia Y, Gao T, Liu Y et al (2020) Zinc isotope revealing zinc's sources and transport processes in karst region. Sci Total Environ 724:138191. https://doi.org/10.1016/j.scitotenv.2020.138191

Zhu X, Tang X, Ng C et al (2025) Biomonitoring xenobiotics in human biospecimens: challenges, advances, and the future of exposome characterization. Rev Environ Contam 263:15. https://doi.org/10.1007/s44169-025-00088-2

Chapter 4
The Impact of Radioactive Materials on Marine Ecosystems: Evaluation and Perspectives

Saloua Mabsor-Zgandaoui, Adil Aknouch, Youssef El-Ouardi, and Khawla Rachmoune

Abstract Radioactive materials from natural, intentional, and accidental sources represent a persistent threat to marine ecosystems and human health. Once introduced into the ocean, radionuclides interact with water, sediments, and organisms, undergoing transport, transformation, and incorporation into food webs. Through bioaccumulation and biomagnification, they can disrupt biodiversity and expose human populations via seafood consumption. This chapter reviews the main sources of marine radioactivity, including nuclear tests, waste disposal, and nuclear accidents, and examines the mechanisms driving dispersion and ecological effects. Risk assessment tools and ecological protection frameworks are presented alongside current advances and challenges. Sustainable management strategies, such as monitoring, waste treatment, and bioremediation, are also discussed to mitigate long-term impacts on marine ecosystems and human health.

S. Mabsor-Zgandaoui
Biology and Health Laboratory, Department of Biology, Faculty of Sciences, Ibn Tofail University, Kenitra, Morocco

A. Aknouch (✉)
LICPM Laboratory, Department of Physics, Faculty of Sciences and Techniques, Sultan Moulay Slimane University, Beni Mellal, Morocco

Higher Institute of Nursing Professions and Health Techniques, Agadir, Morocco
e-mail: adil.aknouch@uit.ac.ma

Y. El-Ouardi
LIMAS Laboratory, Department of Physics, Faculty of Science Dhar El Mahraz, Sidi Mohamed Ben Abdellah University, Fez, Morocco

K. Rachmoune
University of Lyon, Universite Claude Bernard Lyon 1, LAGEPP UMR 5007, Villeurbanne, France

N. Mejjad (ed.), *Nuclear-Based Approaches application for marine pollution assessment and monitoring*, Environmental Science and Engineering,
https://doi.org/10.1007/978-3-032-12197-4_4

Keywords Marine ecosystems · Radionuclides · Radioactivity · Bioaccumulation · Radiological risk · Bioremediation

4.1 Introduction

Since the discovery of radioactivity at the end of the nineteenth century, radioactive materials have become powerful tools in many fields: energy, medicine, industry, research, and the military (Aknouch et al. 2021; El-Ouardi et al. 2023). However, their release into the environment, whether deliberate or accidental, has gradually revealed lasting consequences, particularly in marine environments, which are vast, dynamic and vulnerable.

For a long time, the oceans were seen as virtually infinite reservoirs capable of diluting all forms of pollution. However, inputs of radionuclides from nuclear tests, industrial accidents, controlled or uncontrolled discharges and the dumping of radioactive waste have demonstrated their capacity to disrupt marine biogeochemical balances. In addition to these anthropogenic sources, there is also natural radioactivity, in the form of radionuclides of telluric or cosmic origin, which contributes to the radioactive background in ocean environments (Reichelt-Brushett and Oakes 2023).

Understanding the impact of radioactive materials on marine ecosystems requires an integrated approach, combining geochemistry, ecotoxicology, oceanography and risk assessment. Radionuclides do not simply dissolve in water: they interact with suspended particles, settle in sediments, become part of food webs and can ultimately affect marine organisms and human health via the food chain.

This chapter provides an overall, structured analysis of the problem of radioactive contamination of marine environments, addressing it from five complementary angles. It begins by identifying the sources and mechanisms responsible for this contamination, before examining the physico-chemical behaviour of radionuclides in the ocean environment, taking into account their transport, transformations and interactions with abiotic compartments. The analysis continues with an assessment of the ecological impacts of radioactivity, considered at different biological scales, from the individual to the ecosystem. The chapter then presents the methodologies used to assess radiological risk in the marine environment, highlighting both their limitations and recent advances in the field. Finally, we look at the prospects for sustainable management of these risks, integrating the issues of monitoring, regulation and prevention.

At a time when the oceans are under increasing pressure from climate change, plastic pollution and intensive resource exploitation, the issue of marine radioactivity is often overlooked. The aim of this chapter is to highlight the scientific and societal issues involved, laying the foundations for a precise and up-to-date understanding of the interactions between radionuclides and marine ecosystems.

4.2 Sources of Radioactivity in the Marine Environment

4.2.1 Intentional Sources

4.2.1.1 Nuclear Weapons

At present, nine sovereign states are officially recognised as possessing military nuclear capabilities: Russia, the United States, France, China, the United Kingdom, Israel, Pakistan, India and North Korea. Among them, Russia and the United States hold the vast majority of arsenals, with around 6255 and 5550 nuclear warheads respectively (SIPRI 2021). At the other end of the spectrum, North Korea is estimated to have between 40 and 50 nuclear weapons (SIPRI 2021).

Nuclear testing is the main source of intentional emissions of radioactive materials into the environment, particularly when these weapons are used or tested. In addition to their arsenals, the United States and Russia also possess the world's largest reserves of enriched uranium and separated plutonium, fissile materials essential to the manufacture of nuclear weapons (SIPRI 2021).

Several nuclear test sites have left lasting traces of radioactive contamination, particularly in marine environments. Among the most notable are Novaya Zemlya (Russia), the Marshall Islands (USA), Christmas Island (UK), French Polynesia (France) and Lop Nor (China), where atmospheric and submarine tests were conducted during the Cold War (Livingston and Povinec 2000).

4.2.1.2 Radioactive Waste Management

The first documented case of radioactive waste being dumped at sea dates back to 1946, in the north-east Pacific Ocean, around 80 km off the Californian coast (Dominique 1989). This practice continued for almost 36 years, including the 1982 dumping of radioactive waste more than 550 km from the European continental shelf in the Atlantic Ocean (Dominique 1989). It is estimated that around 63 petabecquerels (PBq)—or 1.7 million curies (MCi)—of radioactive waste from activities linked to research, medicine and the nuclear industry were dumped in this way. This waste, generally packaged in metal drums lined with a mixture of concrete and bitumen, was spread over more than 50 identified sites in the North Atlantic and North Pacific oceans (Dominique 1989).

The human use of radioactive materials inevitably generates waste. As early as 1960, Pritchard pointed out that the production of radioactive waste was an inevitable consequence of the exploitation of atomic energy. He was already calling for rigorous measures to protect human beings and marine resources, and proposing permissible concentrations and specific conditions for the discharge of radioisotopes into the sea (Pritchard 1960). This realisation fuelled important scientific discussions: in 1976, a workshop held at Woods Hole looked into the possibility of storing High-Level Waste (HLW) in the ocean environment, although this option had not been seriously considered at the time (Vilks 1976).

Some of the proposals were bold, even futuristic. It was suggested in particular that plutonium-239, which has a half-life of 25,000 years, should be deposited in the basal rocks of an oceanic plate, close to a subduction zone, so that tectonic movements would eventually carry the waste into the earth's mantle.

Despite the debates and recommendations made in those years, a considerable quantity of radioactive waste was indeed dumped, constituting today an environmental legacy in the form of ageing containers at the bottom of the oceans. This 'legacy waste' raises concerns about the durability of the storage materials and the possibility of long-term radioactive leaks.

The International Atomic Energy Agency (IAEA) has since introduced a standardised classification of radioactive waste, taking into account its physical state (solid, liquid, organic) as well as its radioactive half-life (short or long half-life), in order to provide a framework for its treatment and disposal. The management of radioactive waste resulting from intentional activities (nuclear tests, dumping, submarine or power plant accidents) will have to continue for several generations.

Although the option of disposal at sea was seriously considered in the 1960s and 1970s (Pritchard 1960; Vilks 1976), it has now been ruled out by the international community. However, current sources of man-made radioactivity continue to feed the marine environment, such as global fallout from past atmospheric nuclear tests, or low-level discharges from the normal operation of nuclear power stations and reprocessing plants (Livingston and Povinec 2000).

In addition, the transport of radioactive materials, particularly via international trade in nuclear waste management, represents a significant risk. The majority of low-level waste (LLW) can be processed safely on land, but specialised facilities are needed to process or store intermediate-level waste (ILW) and high-level waste (HLW). These materials are often shipped from the place of production to treatment or storage centres, mainly by sea—logistics which, although regulated, always involve a degree of risk.

4.2.2 Accidental Sources

4.2.2.1 Nuclear Accidents

The mandate of the IAEA, which was established in 1957 and now has 180 member states, is to promote the peaceful uses of nuclear energy while preventing its exploitation for military purposes, including the proliferation of nuclear weapons.

According to its annual report (IAEA 2021), the world's nuclear fleet currently comprises 437 reactors in operation. The trend is towards growth: 56 additional reactors are currently under construction, and according to IAEA projections (2019), nuclear capacity could increase by 12–25% by 2030, and up to 80% by 2050. However, this growth is accompanied by a major challenge: the sustainable management of radioactive waste, a problem for which there is currently no universally recognized solution (Choudri and Baawain 2016).

From an environmental point of view, the location of nuclear power plants is not insignificant. Reactors require large quantities of water for their cooling systems. Seawater, which is available in abundance and does not compete with freshwater resources, is frequently used, which explains the coastal location of many nuclear facilities. With this in mind, some researchers have even proposed the development of floating nuclear reactors, installed on marine platforms (Standring et al. 2009).

One of the main advantages of nuclear power is its ability to produce carbon-free energy. Indeed, countries that rely heavily on nuclear power often have relatively low levels of greenhouse gas emissions. However, the technology remains controversial because of the risks associated with nuclear accidents, whether man-made (Three Mile Island in 1979 and Chernobyl in 1986) or natural (Fukushima, 2011).

These events have left their mark on public opinion and generated persistent mistrust. The long-term consequences for health, ecosystems and food chains, particularly in the marine environment, are fuelling legitimate concerns. In response, numerous scientific studies have looked at the dispersion of radioactivity after major accidents. However, public perception can sometimes be skewed by misinterpretation of the data. For example, a map widely circulated after the Fukushima accident, which was supposed to show the radioactive contamination of the Pacific Ocean, actually came from NOAA and represented the maximum wave heights of the tsunami of 11 March 2011.

Despite this media confusion, several studies have confirmed measurable radioactive contamination in certain local seafood products (Buesseler et al. 2012), as well as in samples from the North Pacific (Azouz and Dulai 2017), with potential transfer via fish migration (Madigan et al. 2012). That said, most of the levels detected remain below the thresholds deemed of concern by health agencies (Buesseler et al. 2012; Azouz and Dulai 2017; Fisher et al. 2013).

4.2.3 Sources Natural

The natural radioactivity of the Earth has played a fundamental role in shaping the planet's evolution. It significantly influenced the differentiation of Earth's internal structure, the formation of an oxygen-rich atmosphere, and the emergence and development of biological life. Radioactive isotopes have been present since the Earth's formation, approximately 4.6 billion years ago. Among the key contributors to the thermal and radiogenic evolution of Earth's matter are potassium-40 (^{40}K, half-life = 1.28 billion years), uranium-235 (^{235}U, half-life = 0.713 billion years), uranium-238 (^{238}U, half-life = 4.47 billion years), and thorium-232 (^{232}Th, half-life = 14.05 billion years) (Choppin et al. 2013). The decay of these long-lived radionuclides generated internal heat, which together with gravitational forces contributed to melting and redistribution of Earth's materials, eventually forming the core, mantle, and crust.

As a result of these primordial processes and ongoing geochemical cycles, natural radioactivity continues to be present in various environmental compartments, including the oceans. The natural radioactivity in the oceans originates from several geochemical and cosmogenic sources. A major contribution comes from the leaching of continental rocks, which releases radionuclides such as ^{238}U, ^{232}Th, and ^{40}K (Abbasi et al. 2022a). These radionuclides enter the marine environment via river discharge and coastal erosion. Among these, ^{40}K, found at relatively stable concentrations in seawater, is a dominant source of natural oceanic radioactivity due to its long half-life and high solubility.

Other important natural isotopes include radium-226 (^{226}Ra) and radium-228 (^{228}Ra), produced in the decay chains of uranium and thorium, which are released from sediments and coastal groundwater through the Submarine Groundwater Discharge (SGD) process. Additionally, carbon-14 (^{14}C), generated naturally in the upper atmosphere by cosmic ray interactions with nitrogen, enters the oceans via air-sea gas exchange and is incorporated into marine organic matter.

Minor yet significant inputs include beryllium-7 (^{7}Be), produced in the stratosphere and deposited onto the oceans through precipitation, as well as radionuclides emitted from deep hydrothermal vents along ocean ridges (Schaar et al. 2025). Furthermore, marine sediments act both as reservoirs and secondary sources of radioactivity, slowly releasing isotopes via diffusion and bioturbation. Together, these processes define the natural radioactive background in marine environments, providing a baseline for assessing anthropogenic radioactive inputs.

4.3 Transport and Dispersion

The partitioning and transport of radionuclides in the marine environment are influenced by the interaction between complex hydrodynamic processes and the physiochemical properties of the radionuclides themselves, which occur in the ocean. These processes determine not only the spatial and temporal distribution of radionuclides, but also their impact on ecology and human health. The key factors are solubility, speciation, particle reactivity, interactions with sediments and the influence of physical oceanographic mechanisms such as currents, turbulence and vertical mixing (IAEA 2024).

4.3.1 Physicochemical Interactions

Radionuclides behave differently in the marine environment, largely determined by their solubility and tendency to associate with particles. Soluble radionuclides such as cesium (^{137}Cs), remain in the dissolved phase, although they may also undergo moderate sorption onto fine particles depending on ambient environmental conditions. However, particle-reactive radionuclides, notably the isotopes of plutonium

(^{239}Pu, ^{240}Pu), thorium (^{230}Th) and americium (^{241}Am), bind readily to suspended particulate matter, in particular organic and mineral particles. These associations facilitate their elimination from the water column by sedimentation, especially in areas of high primary productivity or significant terrestrial inputs (Abbasi et al. 2022b).

Radionuclides behave quite differently in the ocean, where their fate and distribution are largely affected by physical, chemical and biological processes. For example, ^{137}Cs is highly soluble in water but is poorly absorbed by clays and organic particles, which explains why some of it ends up being deposited in sediments (Nakano and Povinec 2003). It is also influenced by environmental factors such as salinity, pH and composition of suspended particles, particularly clay minerals such as montmorillonite. Beyond that, this complexity is further increased when considering radionuclides with variable chemical states. Pu and neptunium, for example, can undergo redox modification in seawater. Each oxidation state has its own characteristic physiochemical properties, which in turn affect their mobility, long-term environmental impact and bioavailability to aquatic life. An accurate understanding of these changing chemical states is therefore necessary to predict how radionuclides accumulate shift, and distribute in ocean environments over time (Ashraf et al. 2014).

4.3.2 Oceanic Transport Mechanisms

Oceanic transport plays a central role in the spread of radionuclides over long distances. The horizontal transport is dominated by surface and intermediate currents, like, the Gulf Stream the Kuroshio Current and the Antarctic Circumpolar Current, that can carry radionuclides across entire ocean basins. Vertical migration is via turbulent mixing, wind-induced upwelling and convective overturning, that can redistribute pollutants into deeper oceans (Ganachaud 2003).

One of the most well-known examples of the transport over long distances of radioisotopes is the dispersion of ^{137}Cs and ^{134}Cs following the Fukushima Daiichi nuclear accident in 2011 is a key example of large-scale radionuclide transport. Both monitoring programmes and ocean circulation models indicated that, within 2 years of the release, measurable concentrations of radionuclides have pervaded the entire North Pacific and reached the west coast of North America. The fact that the evidence backed up earlier predictive models and spurred efforts to develop more advanced data tools, including real-time ocean observations, radionuclides decay rates, and particle behavior, (Vives i Batlle et al. 2018; Lee et al. 2024), further confirmed their validity.

The thermohaline circulation, or ocean conveyor belt, is one of the main drivers of radionuclide transport. This is a massive system connecting surface waters and deep waters, which transports warm waters to the poles and cold waters to the equator. It contributes to maintaining climate regulation and global thermal equilibrium through the redistribution of heat. At the heart of this flow is the meridional overturning circulation, where cold, dense water sinks at high latitudes and slowly rises to the surface, having direct influence on heat transport and the carbon cycle in the ocean (Seidov 2009).

4.3.3 Modeling and Monitoring Approaches for Radionuclide Transport in Marine Systems

A complete grasp of radionuclide behavior in marine environments increasingly relies on the use of coupled numerical models that combine hydrodynamic and biogeochemical processes. These models are designed to stimulate the unique physicochemical properties of radionuclides, including their decay kinetics, particle reactivity, and exchange between dissolved and particulate phases. By incorporating real-time environmental datasets, such as salinity gradients, temperature profiles and current velocity fields, these tools enable the simulation of radionuclide dispersal patterns and accumulation dynamics under both typical and extreme conditions (Harms et al. 2003).

Current advances in ocean observation systems have significantly improved the spatial resolution and accuracy of radionuclide observation. Satellite remote sensing delivers large-scale environmental datasets essential for boundary conditions, while autonomous instruments, such as profiling floats and underwater gliders, supply continuous, in situ measurements from remote or deep ocean regions. These autonomous systems are complemented by direct sampling campaigns carried out aboard oceanographic research vessels and through established coastal monitoring networks, where advanced radiochemical techniques are employed to quantify radionuclide concentrations and determine their chemical speciation (Oros 2025).

New computer tools, including those based on AI and machine learning, could help us better understand and predict complex environmental changes. For dynamic coastal areas, where traditional models often struggle because conditions can change quickly and available data is usually limited, this is especially important. By leveraging large amounts of data, these intelligent algorithms can detect complex patterns and improve our ability to predict how pollutants or sediments move and behave along the coastline (Masria and Abouelsaad 2025).

Global coordination remains essential for effective marine radionuclide monitoring. International institutions such as the IAEA support collaborative efforts by harmonizing data collection protocols, maintaining shared databases, and conducting interlaboratory comparisons to ensure data consistency. These collaborative frameworks enable the identification of long-term trends in marine radioactivity and support the development of early detection systems for anomalous events, including unauthorized discharges, accidental releases, or delayed emissions from legacy contamination sources (IAEA 2024).

4.4 Bioaccumulation and Biomagnification of Radionuclides

Radionuclides introduced into marine ecosystems, whether through atmospheric fallout, direct discharge, or remobilization from sediments are subject not only to physical dispersion but also to biological incorporation. Two key biological processes govern their persistence and transfer within marine food webs: **bioaccumulation**, the uptake and retention of radionuclides by organisms from their surrounding environment, and **biomagnification**, the progressive increase in radionuclide concentration at successive trophic levels (Isibor et al. 2024). Figure 4.1, illustrates how radionuclides build up and get stronger in the sea. The transportation by marine food chains reflects the phenomena of bioaccumulation and biomagnification. As a result, radionuclide concentrations tend to be lowest in algae and crustaceans, but increases in fish and are highest in large predators. This food chain transfer is a critical ecological and health issue, particularly for human consumers of contaminated seafood (Fig. 4.1).

4.4.1 Mechanisms of Bioaccumulation

Bioaccumulation refers to the process by which organisms absorb radionuclides directly from seawater, suspended particles, sediments, or through ingestion of contaminated food. The degree of accumulation depends on several factors, including the chemical behavior of the radionuclide, environmental conditions, like salinity, pH, and species-specific physiology (Isibor et al. 2024). Radionuclides tend to mimic essential biological elements:

^{137}Cs behaves similarly to potassium and is readily absorbed by marine organisms through ion-exchange processes, particularly accumulating in muscle tissues.

^{90}Sr, analogous to calcium, is incorporated into calcified structures such as bones, shells, and otoliths.

^{239}Pu, ^{240}Pu, with high particle reactivity, are less bioavailable in dissolved form but may enter organisms via ingestion of contaminated sediments or particulate-bound matter, especially in benthic feeders like mollusks and polychaetas (Zalewska 2014).

Studies confirm that organisms inhabiting sediment-rich or estuarine zones, where resuspension and particle flux are high, exhibit enhanced radionuclide uptake. The *Concentration Factor* (CF), a standard metric used to quantify this process, varies widely by species and isotope. For example, ^{137}Cs CFs for marine fish can range from 20 to 100, whereas for ^{241}Am in benthic invertebrates, CFs can exceed 10,000, indicating strong particle association and sediment interaction (Abdelkarim and Imam 2024).

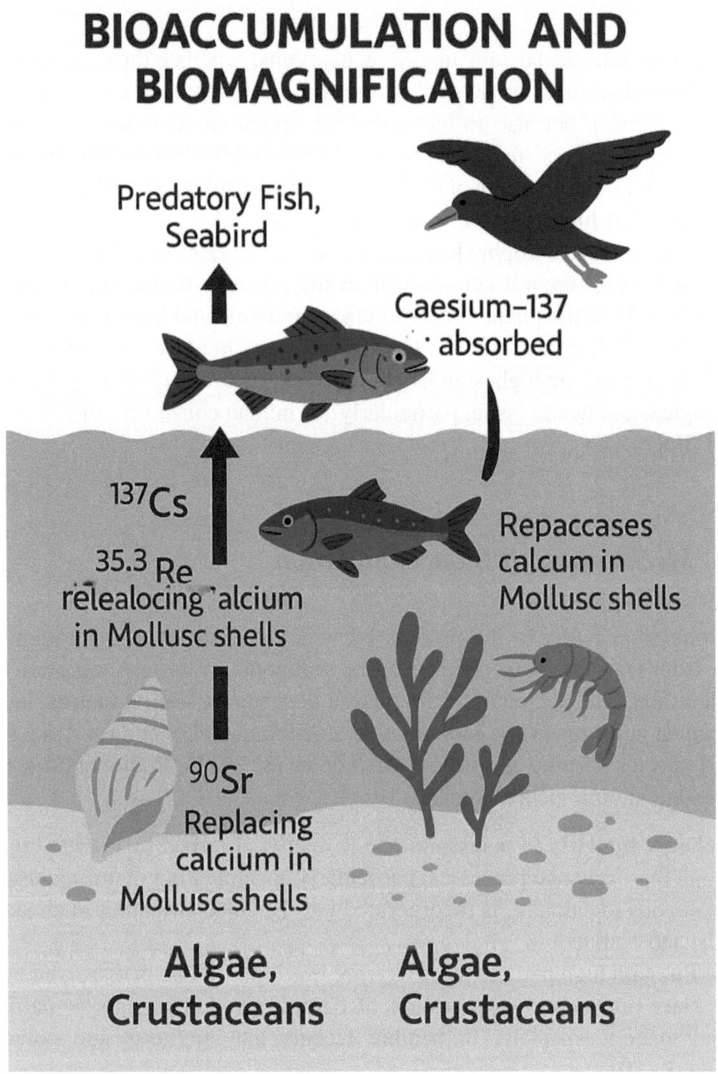

Fig. 4.1 Bioaccumulation and biomagnification of radionuclides in the marine environment. Radionuclides such as cesium-137 and strontium-90 penetrate the food chain via primary producers and small invertebrates (mollusks, crustaceans and algae), where they replace essential elements such as calcium and potassium. Gradually, these radionuclides accumulate in fish and end up concentrated in large predators, resulting in higher concentrations at successive trophic levels

4.4.2 Trophic Transfer and Biomagnification

While bioaccumulation describes the initial uptake, **biomagnification** refers to the increase in radionuclide concentrations as they ascend the food chain from phytoplankton to zooplankton, to fish, and lastly to apex predators species such as marine mammals, seabirds, and humans (Saidon et al. 2024).

Radionuclides differ in their tendency to bio-magnify. ^{137}Cs, due to its biological retention and assimilation efficiency, shows measurable biomagnification across multiple trophic levels. For instance, higher concentrations of ^{137}Cs have been observed in piscivorous fish like tuna and swordfish, especially following nuclear accidents such as Fukushima. On the other hand, ^{90}Sr generally does not exhibit significant biomagnification because of its low assimilation efficiency and relatively fast elimination from higher trophic organisms. The extent of biomagnification is influenced by:

Trophic level position of the organism.
Lipid solubility and binding affinity of the radionuclide.
Dietary habits and metabolism.
Biological half-life of the radionuclide in the organism.

Ecological studies confirm that long-lived radionuclides can persist in food webs for extended periods. For example, Pu and Am can be retained in digestive glands or soft tissues of benthic feeders for months to years, representing a continuous source of internal exposure (Saidon et al. 2024).

4.4.3 Ecological and Human Health Issues

The accumulation of radionuclides within marine food webs raises concerns about sub-lethal effects on marine organisms and potential health risks to humans who consume contaminated seafood. Long-term exposure to low levels of radionuclides, particularly alpha emitters like ^{239}Pu or beta emitters like ^{90}Sr, can pose carcinogenic or mutagenic risks.

Monitoring programs and risk assessments must therefore consider not only total radionuclide concentrations but also **bioavailability**, **trophic transfer rates**, and **species-specific accumulation profiles**. International guidelines, including those from the IAEA and FAO, emphasize the need for continuous surveillance of radionuclides in seafood, especially in regions influenced by historical nuclear tests, nuclear facility discharges, or recent incidents such as the Fukushima Daiichi disaster (IAEA 2024).

4.5 Impact of Radioactive Material on Marine Ecosystems

The presence of radioactive materials in marine ecosystems poses a major threat to both biodiversity and human health. Ecological Impacts Radioactive contamination can severely harm marine life, from plankton to large marine mammals. In the marine ecosystem, radionuclides can accumulate in the food chain, causing mutations, physiological pathologies or even the death of certain species. This bioaccumulation also represents a direct threat to human beings, in particular in cases of consumption of contaminated fish, or contaminated seawater in some bathing areas. These activities can involve exposing populations to levels of radiation which could lead to harmful health effects such as cancer, genetic damage or long-term diseases.

Sources of radioactive pollution in marine environments are numerous and often linked to large-scale human activities. One of the earliest historical sources was nuclear weapons testing conducted in the oceans, particularly between the 1940s and 1980s, which released significant amounts of radionuclides into the water. Another major source is the dumping of radioactive waste and other nuclear materials directly into the seas, a practice that was common until the late twentieth century, especially in the North Atlantic and Arctic regions. Additionally, radioactive effluents from nuclear facilities (power plants, reprocessing plants, laboratories) are regularly discharged into coastal areas, often through authorized procedures that nevertheless have long-term environmental impacts. Two major events have also significantly contributed to marine radioactive contamination: the Chernobyl disaster in April 1986, whose atmospheric fallout reached several European seas, and the Fukushima Daiichi catastrophe in March 2011, which led to a massive release of radioactive materials into the Pacific Ocean. These incidents have had significant and long-term ecological consequences, highlighting the vulnerability of marine environments to this type of pollution (Nies 2018).

Radionuclides in seawater are harmful to living beings and human health. These materials emit ionizing radiation that affects marine organisms, often causing cellular destruction that is usually irreparable. This radiation primarily affects DNA, leading to genetic mutations, strand breaks, or alterations in gene behavior. Ionizing radiation can also affect the water molecule inside the cell by producing highly reactive free radicals that disrupt normal cellular mechanisms, leading to metabolic changes, premature apoptosis, or even sterility Mabsor-Zgandaoui et al. (2025a). In humans, the bioconcentration of radionuclides through the marine food web represents a major risk: consumption of contaminated seafood can expose internal organs to ionizing radiation, leading to side effects such as cancer, birth defects, or even long-term degenerative changes Mabsor-Zgandaoui et al. (2025b).

In a marine environment, various factors determine the risk associated with radionuclides. The nature of the radionuclide is a determining factor. Indeed, some isotopes, such as cesium-137 or strontium-90, mix easily with water and are readily absorbed by living organisms; others, such as plutonium-239 have a high affinity for sediments. Half-life is also crucial; the longer it is, the longer the contamination persists. The concentration of radionuclides in the marine environment is a crucial element that determines the intensity of exposure. Furthermore, according to the IAEA, the concentration factor (CF) also contributes to bioaccumulation, as it demonstrates the ability of an organism to accumulate a radionuclide present in seawater. This depends on the marine specificity, the radionuclide, and the physicochemical properties of the environment. In fact, some radionuclides show a greater tendency to bioaccumulate in mollusks and crustaceans than in fish. Another essential factor in assessing the mobility and fixation of radionuclides in the marine environment is the sediment distribution coefficient (Kd); high Kd values mean that the radionuclide is strongly retained by the sediment, showing limited immediate biological availability, but high long-term persistence. These factors show that the ecotoxicological effects of radionuclides cannot be assessed as constant values, but depend on a combination of nature and substance, environmental behavior, and specific interactions with the exposed species (IAEA 2010; Kırıs and Baltas 2019).

The Fukushima Daiichi nuclear disaster caused the release of radionuclides, particularly Cesium-134, Cesium-137, Strontium-90, and Tritium into the Pacific Ocean. These radioactive containing compounds, together with radionuclides of long half-lives, entered the food chain through bioaccumulation and biomagnification, causing the pollution of various marine organisms, from lower to higher trophic individuals, including commercially important species. Some types of catches were regulated as high concentrations of cesium were detected on some species. This exposure had immediate implications for food safety as chronic exposure of food consumers to low dosage ionizing radiation is believed to increase the risk of cancer and other long-term health problems. The presence of some radionuclides in marine environments more than a decade after the event, is evidence of the magnitude and the duration of pollution (Buesseler et al. 2012; Lu et al. 2021).

In 2014, Janovics et al. (2014) conducted a study in the vicinity of the Paks nuclear power plant in Hungary measured concentrations of tritium, cobalt-60, and cesium-137 in the water, sediment and aquatic biota in the Danube River. Analysis results showed the small increase of tritium concentration in downstream of the discharge canal in the right middle level and detection of cobalt-60 in composition of sediments. In the case of molluscs, a relatively high specific activity of OBT (Organically Bound Tritium) was seen in individuals collected downstream, suggesting marked biological uptake. While the estimated doses are low measured effects of legally permitted discharges of radionuclides to the aquatic environment are demonstrated in this study and emphasize the necessity of persistent monitoring of the environment.

With regard to the radioactive contamination, a second study a second study by Antovic and Antovic (2011) was conducted on the accumulation of cesium-137 and radium-226 by the species Chelon labrosus in the southern part of Adriatic Sea. Cesium was present in 5 of 22 fish with a range in activity from 0.32 to3.7 Bq/kg in muscle tissues and radium was detected in all fish with a range from 3.8 to 25 Bq/kg in skeleton. From seawater, average concentration factors were 47 for cesium in muscle and 130 for radium in bone. The study calculated that eating this species annually would yield an effective dose of 4.1 μSv/year for an adult. This provides a vivid example of the capability of some species to accumulate radionuclides and consequent possible human food chain implications.

National and international measures have been developed to reduce the risks posed by radioactivity in marine environments, limiting the potential impact of radionuclides on human health and the environment. One of the most important actions is the monitoring of the concentration of radionuclides in seawater, sediments, and biota. It should be noted that for specific radionuclides, the IAEA provides recommendations on concentration limits; for expanded inspection, e.g., the reference level of ^{137}Cs in drinking water is 10 Bq/L based on IAEA recommendation, or the indicative national emergency level for a food product (specifically fishery products) the reference level is about 1000 Bq/kg for ^{137}Cs. These references are then used to make fishery area closures, to ban or to restrict fishing or the consumption of marine products contaminated with radionuclides from the sea (Nies 2018).

Controlled discharges of radioactive substances emanating from nuclear sources are also supervised by tough authorizations using same site dose assessments, based action levels of 1 mSv/year for public exposure due to industrial activities. Furthermore, application of marine dispersion models can be employed to forecast the spread of radionuclides in the environment and estimate the vulnerable sites. The IAEA also encourages the adoption of comprehensive strategies, including the Environmental Risk Assessment and Management of Ionizing Contaminants (ERICA) framework, to assess potential impacts on marine life. This collaborative international approach facilitates the harmonization of assessment methods and transparency of information. It also promotes the implementation of common response and prevention protocols for rigorous incident management and the long-term protection of marine environments.

4.6 Risk Evaluation and Methodologies

Marine evaluating radiological risks requires an integrated multidisciplinary address that integrates ecological knowledge and modelling and radiation protection principles. Following the entry of radioactive materials into marine systems through discharges, atmospheric deposition, or mobilized sediments, their possible impacts on aquatic organisms and ecosystems need to be fully considered. There are two essential pillars of risk assessment involved, radioecological transfer models, designed to simulate the movement and accumulation of radionuclides in different environmental compartments, and radiation protection approaches, which provide regulatory frameworks and ecological thresholds to safeguard the well-being of marine ecosystems (Huang et al. 2025).

4.6.1 *Radioecological Transfer Models*

Radiological-ecological models are essential for predicting what happens in marine ecosystems with regard to radionuclides. These models are intended to model the transport, transformation, and bioaccumulation of radioactive substances in different compartments, including water, sediments, and living organisms. Their usefulness is not restricted to emergency scenarios; they are equally important for routine monitoring, environmental impact assessments, and long-term decontamination planning (Radakovitch et al. 2024).

4.6.1.1 Principles and Structure of the Transfer Model

The vast majority of radioecological transfer models are modular models, which separate the environment is divided into discrete elements such as the water column, the seabed, plankton, and fish. These representations mathematically depict how radionuclides move from one compartment to another, accounting for biogeochemical processes such as sorption, desorption, sedimentation, and biological uptake, in addition to taking physical transport mechanisms such as advection and diffusion, as well as (Abbasi et al. 2022b; Harms et al. 2003).

One of the key features of these methods is the use of parameters such as transfer coefficients and distribution parameters. For instance, the Distribution Coefficient (Kd) expresses the ratio of the concentration of a particular radionuclide on particles relative to that in the solution phase. The FC expresses the absorption of the radionuclide in the plant and animal material relative to its content in ocean water, whereas the Trophic Transfer Factors reflect the efficiency of radionuclide delivery via the food web. Early models operated on the principle of assuming stable conditions, but modern methods incorporate elements of dynamics to account for time-dependent behaviors such as seasonal changes, episodic releases, and the life cycles of organisms. Models can simulate short- and long-term scenarios, providing a more realistic representation of radionuclide distribution and accumulation (Kumar et al. 2020).

4.6.1.2 Spatial Resolution and Dynamic Modeling

Increasingly, programs are modelling spatial details more accurately by including ocean currents and additional hydrodynamic data. This makes it possible to stimulate the dispersion of radionuclides over large spatial domains, ranging from regional areas to global oceans. These types of models are most useful in estimating contamination distributions after accidents or ongoing releases. Important examples include the IAEA's EMRAS/MODARIA models, used to stimulate environmental transfers between ecosystems; the ERICA tool, which combines environmental concentration data with dose calculations for reference organisms; and PREDO, which estimates radiation doses associated with seafood consumption (IAEA 2024).

Their accuracy depends heavily on the quality and amount of environmental and biological data that are available. Critical inputs include oceanographic conditions like currents and temperature, food web processes, species-specific uptake factors, and water and sediment chemistry. Because of natural variability, simplifications, and limits of measurement, uncertainties always exist but are kept to a minimum using sensitivity analysis and probabilistic modeling.

The use of radioecological models is widespread. They are employed to assess impact in order to estimate the levels of exposure of organisms in the vicinity of release sites or contaminated areas, to plan remediation measures in order to predict the long-term behavior of radionuclides and guide cleanup efforts, and to support regulation in order to promote environmental protection.

4.6.2 Radiation Protection Approaches for Marine Environments

The protection of marine ecosystems from radiation requires complete frameworks that go beyond human health to also protect marine species and habitats. In the last two decades, the focus has shifted from protecting only people to also protecting the health of marine plants and animals (Lagaki et al. 2024).

4.6.2.1 From a Human-Centered to an Eco-Centered Approach

Radiation protection has historically focused primarily on protecting people, and environmental monitoring has mostly been done to make sure that the radiation exposures to humans stay within safe limits. This method, however, frequently disregards a crucial fact that dosages deemed safe for people might nonetheless be detrimental to marine life. The International Commission on Radiological Protection (ICRP) expanded its viewpoint by including ecological considerations in its recommendations. A major milestone was ICRP Publication 108, which adopted a specific framework for safeguarding non-human species by addressing environmental exposure to ionizing radiation directly, was a significant turning point (ICRP 2008).

The ICRP's ecological approach of the ICP is based on the concept of reference animals and plants (RAP). These are representative species chosen to reflect key ecological functions and exposure pathways in various ecosystems. For instance, RAPs in marine environment include bivalve mollusks, known for their ability to filter and to interact with sediments; crustaceans such as crabs, which interact with both sediment and water; flatfish, which live on the bottom and are exposed to contaminated sediments; and algae, which are primary producers sensitive to contaminants in water. These models are used to develop derived reference levels (DRLs), which show the points at which dose ranges at which ecological effects on populations are more likely to occur (ICRP 2008).

4.6.2.2 Reference Animals and Plants

For the purpose of evaluating ecological risks from radionuclides, a number of assessment and policy frameworks have been developed, building on this paradigm. For example, the ERICA tool offers an integrated approach that combines environmental data and dose modeling to deliver a thorough risk assessment. While the R&D128 tool is used in UK for initial screening of radiation effects, in the United States, RESRAD-BIOTA is commonly utilized to assess potential radiological impacts in aquatic and terrestrial environments. The UK's tool is also used for initial assessments of radiation effects. These tools are essential resources for decision-makers, as they help them determine whether the presence of radionuclides may pose a risk to aquatic organisms and whether protective measures are necessary (Prlić et al. 2017).

The Reference Animal and Plant (RAP) technique has been adopted by international organizations including the IAEA and the United Nations Scientific Committee on the Effects of Atomic Radiation (UNSCEAR) as a component of their environmental protection plans. Efforts like the IAEA's MODARIA II initiative aim to enhance the consistency and quality of data acquisition, refine the accuracy of predictive power models, and strengthen technical capabilities in different countries. International and regional collaboration is essential, not only to share knowledge and methodology, but also to address transboundary radiological threats and ensure effective ocean ecosystem protection (Prlić et al. 2017).

4.7 Outlook and Sustainable Management

4.7.1 Surveillance and Research

Environment monitoring of radionuclides in marine ecosystems is a strategic priority for anticipating risks related to radioactive contamination. Radionuclides such as ^{137}Cs, ^{90}Sr, $^{239}Pu/^{240}Pu$, and ^{3}H are environmentally persistent and have a high potential to bioaccumulate. Such elements can be transferred throughout marine food chains, impacting benthic and pelagic organisms and, ultimately, human populations that consume seafood. In order to effectively detect their presence, it is essential to strengthen monitoring programs using advanced technologies such as high-resolution gamma spectrometry, alpha spectrometry, and inductively coupled plasma mass spectrometry (ICP-MS/MS), which enable accurate trace detection (Gwynn et al. 2024).

The nuclear accident at Fukushima in 2011 clearly illustrates the importance of these devices. Indeed, high levels of ^{137}Cs were detected in various marine species, with concentrations sometimes exceeding 1000 Bq/kg, well above the admissible thresholds for human consumption (Ota et al. 2023). It highlighted the need to couple environmental, biological and hydrodynamic data into an integrated approach to monitoring.

In addition to detection, in-depth research into the mechanisms of radionuclide toxicity is essential. On a cellular and molecular scale, chronic exposure induces effects such as the production of reactive oxygen species (ROS), DNA damage, cell cycle disruption and apoptosis (Khazaei Monfared et al. 2023). Intracellular transport of radioactive ions is actively mediated by specific membrane transporters, such as the ZIP, NRAMP and ABC proteins. The emergence of "multi-omics" approaches—including transcriptomics, proteomics and metabolomics—is furthering our understanding of biological responses to radionuclides. Such techniques have enabled the identification of biomarkers of radiotoxicity, including metallothionein genes, antioxidant enzymes such as superoxide dismutase (SOD) and catalase (CAT), and DNA repair proteins (Yadav et al. 2023).

4.7.2 *Reducing Waste*

Limited release of radioactive elements into the aquatic system is an essential lever for controlling the long-term effects on marine ecosystems. This means strictly implementing progressive effluent treatment techniques, complying with international criteria and extending our nuclear waste management strategy (Yue and Yang 2024).

Potential technologies to investigate include the adsorption process, mediated by natural or synthetic materials, such as zeolites, modified clays, biochars and MOFs, which are capable of selectively adsorbing radioactive ions due to ion exchange or surface (Al-Hazmi et al. 2024). Additional interactions are the most promising. Ion exchange resins are also widely used in nuclear power plants for the removal of ^{137}Cs and ^{90}Sr from waste streams. At the nuclear power plant in Fukushima, the simplified active water recovery system (SARRY) plays a crucial role in lowering the concentration of cesium in the cooling water (Lehto et al. 2019).

In addition, dissolved radionuclides can be effectively removed by nanofiltration and reverse osmosis, while the use of chemical precipitation to immobilize metal isotopes is becoming a technique of choice using specific agents such as potassium ferrocyanide.

International standards apply to these activities. The ICRP and the IAEA have recommended annual exposure limits for the general public of less than 1 mSv. UNSCEAR also stresses the importance of taking into account effects on non-human species and ecosystem services. In this context, the long-term management of radioactive waste also requires safe conditioning, in particular vitrification and deep geological disposal.

4.7.3 Recovery and Restoration

Radioactive environmental contamination of the marine environment can only be prevented by taking steps to restore the affected ecosystem to its original state. Among these approaches, bioremediation is one of the most interesting. It relies on the ability of organisms to eliminate (i.e. retain, sequester or transform) radionuclides (Thakare et al. 2021). The microalgae Chlorella vulgaris and Scenedesmus obliquus, and the cyanobacteria Spirulina platensis, have been shown to have a high potential for absorbing uranium, cobalt and cesium, due to the presence of functional groups (carboxyl, phosphate, hydroxyl) on their cell surfaces (Lehto et al. 2019).

In addition, certain radiation-resistant bacteria (notably the genera Shewanella and Geobacter) can partially or totally reduce hexavalent uranium (U(VI)) to its tetravalent form (U(IV)), which is poorly soluble and therefore less mobile in nature (Coral et al. 2022). These properties offer a robust pathway for in situ use in coastal or polluted sediments.

Research suggests that although phytoremediation is not common practice in the marine environment, it is a promising option as a biological remediation method. In particular, halophytes such as Salicornia europaea could be particularly interesting for the uptake and bioaccumulation of radionuclides in saline environments. Such salt-tolerant species could be introduced to constructed marshes or sea buffer zones, serving as a type of natural biological filter. An environmental remedy for the restriction of the migration of the radionuclide into vulnerable aquatic environments (Mohammadzadeh and Hajiboland 2022).

Environmental restoration of marine ecosystems involves concrete actions such as replanting seagrass beds (Posidonia oceanica), reintroducing filter-feeding species (mussels, oysters, sea cucumbers), or installing artificial reefs to encourage benthic recolonization. These measures must be accompanied by rigorous post-restoration monitoring using bioindicators (copepods, bivalves) to verify the effective reduction in radioactivity and the stability of recolonized communities (Lunt et al. 2017).

4.8 Conclusion

The presence of radioactive substances in the marine environment poses a persistent threat to ecosystems and human health, whether they originate from natural processes, deliberate activities, or accidental releases. Radionuclides are not passive pollutants; they disperse, accumulate and integrate into marine food webs, disrupting biodiversity and exposing human populations through the consumption of seafood.

Addressing these challenges requires an integrated scientific approach that combines oceanography, ecotoxicology, radiobiology and advanced modelling. It also requires radiological protection frameworks that extend beyond human safety to encompass the preservation of entire ecosystems.

Looking ahead, the sustainable management of marine radioactivity will require rigorous environmental monitoring, predictive modelling and innovative mitigation strategies, such as advanced waste treatment, bioremediation and ecological restoration.

In an era of climate change and increasing human-caused pressures, marine radioactivity can no longer be considered a marginal issue. It is a central challenge for environmental protection and global sustainability, requiring international collaboration and long-term scientific commitment.

References

Abbasi A, Algethami M, Bawazeer O, Zakaly HMH (2022a) Distribution of natural and anthropogenic radionuclides and associated radiation indices in the Southwestern coastline of Caspian Sea. Mar Pollut Bull 178:113593. https://doi.org/10.1016/j.marpolbul.2022.113593

Abbasi A, Zakaly HMH, Algethami M, Abdel-Hafez SH (2022b) Radiological risk assessment of natural radionuclides in the marine ecosystem of the Northwest Mediterranean Sea. Int J Radiat Biol 98:205–211. https://doi.org/10.1080/09553002.2022.2020359

Abdelkarim MS, Imam N (2024) Radiation hazards and extremophiles bioaccumulation of radionuclides from hypersaline lakes and hot springs. Int J Environ Sci Technol 21:3021–3036. https://doi.org/10.1007/s13762-023-05154-7

Aknouch A, El-Ouardi Y, Hamroud L, Sebihi R, Mouhib M, Yjjou M, Didi A, Choukri A (2021) A Monte Carlo study to investigate the feasibility to use the Moroccan panoramic irradiator in sterile insect technique programs. Radiat Environ Biophys 60:673–679. https://doi.org/10.1007/s00411-021-00934-6

Al-Hazmi GH, Albedair LA, Alatawi RAS, Alsuhaibani AM, Bukhari AAH, El-Bindary AA (2024) Effective synthesis and characterization of citric acid cross-linking of modified ferrous metal-organic framework and chitosan nanocomposite sponge for Th(IV) elimination: adsorption isotherms, kinetic analysis, and optimization by Box-Behnken design. Int J Biol Macromol 281:136194. https://doi.org/10.1016/j.ijbiomac.2024.136194

Antovic I, Antovic NM (2011) Determination of concentration factors for Cs-137 and Ra-226 in the mullet species *Chelon labrosus* (Mugilidae) from the South Adriatic Sea. J Environ Radioact 102:713–717. https://doi.org/10.1016/j.jenvrad.2011.04.006

Ashraf MA, Akib S, Maah MJ, Yusoff I, Balkhair KS (2014) Cesium-137: radio-chemistry, fate, and transport, remediation, and future concerns. Crit Rev Environ Sci Technol. https://doi.org/10.1080/10643389.2013.790753

Azouz HR, Dulai H (2017) In the wake of Fukushima: radiocesium inventories of selected North Pacific fish. Pac Sci 71:107–115. https://doi.org/10.2984/71.2.1

Buesseler KO, Jayne SR, Fisher NS, Rypina II, Baumann H, Baumann Z, Breier CF, Douglass EM, George J, Macdonald AM, Miyamoto H, Nishikawa J, Pike SM, Yoshida S (2012) Fukushima-derived radionuclides in the ocean and biota off Japan. Proc Natl Acad Sci USA 109:5984–5988. https://doi.org/10.1073/pnas.1120794109

Choppin G, Liljenzin J, Rydberg J, Ekberg C (2013) Radiochemistry and nuclear chemistry.4th edn. Elsevier. https://doi.org/10.1016/C2011-0-07260-5

Choudri BS, Baawain M (2016) Radioactive wastes. Water Environ Res 88:1486–1503. https://doi.org/10.2175/106143016X14696400495299

Coral T, Placko A-L, Beaufort D, Tertre E, Bernier-Latmani R, Descostes M, De Boissezon H, Guillon S, Rossi P (2022) Biostimulation as a sustainable solution for acid neutralization and

uranium immobilization post acidic in-situ recovery. Sci Total Environ 822:153597. https://doi.org/10.1016/j.scitotenv.2022.153597
Dominique C (1989) Ocean disposal of radioactive waste. Status Rep 31:47–50
El-Ouardi Y, Aknouch A, Dadouch A, Mouhib M, Maghnouj A, Benmessaoud M, Yjjou M (2023) Control of transit doses by Monte Carlo simulation inside an ionization casemate housing of a ^{60}Co gamma irradiator. Radiat Phys Chem 206:110776. https://doi.org/10.1016/j.radphyschem.2023.110776
Fisher NS, Beaugelin-Seiller K, Hinton TG, Baumann Z, Madigan DJ, Garnier-Laplace J (2013) Evaluation of radiation doses and associated risk from the Fukushima nuclear accident to marine biota and human consumers of seafood. Proc Natl Acad Sci USA 110:10670–10675. https://doi.org/10.1073/pnas.1221834110
Ganachaud A (2003) Large-scale mass transports, water mass formation, and diffusivities estimated from World Ocean Circulation Experiment (WOCE) hydrographic data. J Geophys Res Oceans 108:3213. https://doi.org/10.1029/2002jc001565
Gwynn JP, Hatje V, Casacuberta N, Sarin M, Osvath I (2024) The effect of climate change on sources of radionuclides to the marine environment. Commun Earth Environ 5:135. https://doi.org/10.1038/s43247-024-01241-w
Harms IH, Karcher MJ, Burchard H (2003) Chapter 3: Modelling radioactivity in the marine environment: the application of hydrodynamic circulation models for simulating oceanic dispersion of radioactivity. In: Radioactivity in the environment, vol 4, pp 55–85. https://doi.org/10.1016/S1569-4860(03)80059-1
Huang WH, Chen ZM, Chen TC, Yeh YL (2025) Assessing radiological risks of natural radionuclides on sustainable campus environment. Sustainability (Switzerland) 17:691. https://doi.org/10.3390/su17020691
IAEA (2010) Handbook of parameter values for the prediction of radionuclide transfer in terrestrial and freshwater environments. International Atomic Energy Agency
IAEA (2021) IAEA annual report for 2021. https://www.iaea.org/publications/reports/annual-report-2021. Accessed 26 Jul 2025
IAEA (2024) Enhancement of modelling approaches for the assessment of radionuclide transfer in the marine environment. International Atomic Energy Agency, p 10.61092/IAEA.PJUC-AHA9
ICRP (2008) Environmental protection: the concept and use of reference animals and plants. Ann ICRP. https://doi.org/10.1016/j.icrp.2009.04.002
Isibor PO, Kayode-Edwards II, Imoobe TO, Binuyo OA, Akinsanya B, Akinhanmi FO (2024) Bioaccumulation and biomagnification in Arctic food webs. In: Arctic marine ecotoxicology: climate change, pollutants, and their far-reaching effects. Springer Nature, pp 161–178. https://doi.org/10.1007/978-3-031-73584-4_7
Janovics R, Bihari Á, Papp L, Dezső Z, Major Z, Sárkány KE, Bujtás T, Veres M, Palcsu L (2014) Monitoring of tritium, ^{60}Co and ^{137}Cs in the vicinity of the warm water outlet of The Paks Nuclear Power Plant, Hungary. J Environ Radioact 128:20–26. https://doi.org/10.1016/j.jenvrad.2013.10.023
Khazaei Monfared Y, Heidari P, Klempner SJ, Mahmood U, Parikh AR, Hong TS, Strickland MR, Esfahani SA (2023) DNA damage by radiopharmaceuticals and mechanisms of cellular repair. Pharmaceutics 15:2761. https://doi.org/10.3390/pharmaceutics15122761
Kırıs E, Baltas H (2019) Sediment distribution coefficients (Kd) and bioaccumulation factors (BAF) in biota for natural radionuclides in eastern Black Sea coast of Turkey. Microchem J 149:104044. https://doi.org/10.1016/j.microc.2019.104044
Kumar A, Rout S, Pulhani V, Kumar AV (2020) A review on distribution coefficient (Kd) of some selected radionuclides in soil/sediment over the last three decades. J Radioanal Nucl Chem. https://doi.org/10.1007/s10967-019-06930-x
Lagaki V, Siltzovalis G, Madesis I, Vasileiou P, Mertzimekis TJ (2024) Monitoring of the radioactivity in the marine environment: a White Paper-Part I. Nucl Energy Technol 10:179–187. https://doi.org/10.3897/nucet.10.133710

Lee ST, Cho YK, Jung J, Chae S (2024) Surface and subsurface dispersal of radioactive materials from Fukushima by subpolar gyre and intermediate waters in the North Pacific. Sci Rep 14. https://doi.org/10.1038/s41598-024-55328-7

Lehto J, Koivula R, Leinonen H, Tusa E, Harjula R (2019) Removal of radionuclides from Fukushima Daiichi waste effluents. Sep Purif Rev 48:122–142. https://doi.org/10.1080/15422119.2018.1549567

Livingston HD, Povinec PP (2000) Anthropogenic marine radioactivity. Ocean Coast Manag 43:689–712. https://doi.org/10.1016/S0964-5691(00)00054-5

Lu Y, Yuan J, Du D, Sun B, Yi X (2021) Monitoring long-term ecological impacts from release of Fukushima radiation water into ocean. Geogr Sustainability 2:95–98. https://doi.org/10.1016/j.geosus.2021.04.002

Lunt J, Reustle J, Smee DL (2017) Wave energy and flow reduce the abundance and size of benthic species on oyster reefs. Mar Ecol Prog Ser 569:25–36. https://doi.org/10.3354/meps12075

Mabsor-Zgandaoui, S. et al. (2025a) Ionising and Ultraviolet Radiation Induce the Secretion of Melanin in Tetrahymena Pyriformis. *International Journal of Environmental Studies,* 82 (1): 1–14. https://doi.org/10.1080/00207233.2024.2412455

Mabsor-Zgandaoui, S. et al. (2025b) "Effect of Hydrogen Peroxide Pre-Treatment on Enhancing the Cellular Resistance of Tetrahymena Pyriformis Exposed to Ionizing Radiation." *Toxicological & Environmental Chemistry,* 107 (2): 294–309. https://doi.org/10.1080/02772248.2025.2460804

Madigan DJ, Baumann Z, Fisher NS (2012) Pacific bluefin tuna transport Fukushima-derived radionuclides from Japan to California. Proc Natl Acad Sci U S A 109:9483–9486. https://doi.org/10.1073/pnas.1204859109

Masria A, Abouelsaad O (2025) Artificial intelligence applications in coastal engineering and its challenges—a review. Cont Shelf Res. https://doi.org/10.1016/j.csr.2025.105425

Mohammadzadeh P, Hajiboland R (2022) Phytoremediation of nitrate contamination using two halophytic species, Portulaca oleracea and Salicornia europaea. Environ Sci Pollut Res Int 29:46127–46144. https://doi.org/10.1007/s11356-022-19139-5

Nakano M, Povinec PP (2003) Oceanic general circulation model for the assessment of the distribution of ^{137}Cs in the world ocean. Deep-Sea Res II Top Stud Oceanogr 50:2803–2816. https://doi.org/10.1016/S0967-0645(03)00149-8

Nies H (2018) Origin and management of radioactive substances in the marine environment. In: Salomon M, Markus T (eds) Handbook on marine environment protection: science, impacts and sustainable management. Springer International Publishing, Cham, pp 733–753. https://doi.org/10.1007/978-3-319-60156-4_38

Oros A (2025) Bioaccumulation and trophic transfer of heavy metals in marine fish: ecological and ecosystem-level impacts. J Xenobiot. https://doi.org/10.3390/jox15020059

Ota M, Takahara S, Yoshimura K, Nagakubo A, Hirouchi J, Hayashi N, Abe T, Funaki H, Nagai H (2023) Soil dust and bioaerosols as potential sources for resuspended ^{137}Cs occurring near the Fukushima Dai-ichi nuclear power plant. J Environ Radioact 264:107198. https://doi.org/10.1016/j.jenvrad.2023.107198

Pritchard DW (1960) The application of existing oceanographic knowledge to the problem of radioactive waste disposal into the sea. In: Disposal of radioactive wastes. Vol. II. Proceedings of the scientific conference on the disposal of radioactive wastes, pp 229–248

Prlić I, Mostečak A, Mihić MS, Veinović Ž, Pavelić L (2017) Radiological risk assessment: an overview of the ERICA integrated approach and the ERICA tool use. Arch Ind Hyg Toxicol. https://doi.org/10.1515/aiht-2017-68-3020

Radakovitch O, Boyer P, Charmasson S, Duffa C, Durand H, Guillory F, Lepage H, Seyer T (2024) Modelling the transfer of radionuclides in aquatic environments. IRSN

Reichelt-Brushett A, Oakes JM (2023) Radioactivity. In: Reichelt-Brushett A (ed) Marine pollution—monitoring, management and mitigation. Springer Nature Switzerland, Cham, pp 229–245. https://doi.org/10.1007/978-3-031-10127-4_10

Saidon NB, Szabó R, Budai P, Lehel J (2024) Trophic transfer and biomagnification potential of environmental contaminants (heavy metals) in aquatic ecosystems. Environ Pollut. https://doi.org/10.1016/j.envpol.2023.122815

Schaar K, Spiegl T, Sato T, Langematz U (2025) The impact of ENSO on near-surface Beryllium-7. J Environ Radioact 282:107592. https://doi.org/10.1016/j.jenvrad.2024.107592

Seidov D (2009) Heat transport, oceanic and atmospheric. In: Encyclopedia of earth sciences series. Springer Netherlands, pp 1246–1249. https://doi.org/10.1007/978-1-4020-4411-3_102

SIPRI (ed) (2021) SIPRI yearbook 2021: armaments, disarmament and international security. Oxford University Press, Oxford, UK

Standring WJF, Dowdall M, Amundsen I, Strand P (2009) Floating nuclear power plants: potential implications for radioactive pollution of the northern marine environment. Mar Pollut Bull 58:174–178. https://doi.org/10.1016/j.marpolbul.2008.11.025

Thakare M, Sarma H Datar S, Roy A, Pawar P, Gupta, K, Pandit, S, Prasad R. (2021). Understanding the holistic approach to plant-microbe remediation technologies for removing heavy metals and radionuclides from soil. Current Research in Biotechnology. 3, 84–98. https://doi.org/10.1016/j.crbiot.2021.02.004

Vilks G (1976) The disposal of high level nuclear waste in the oceans. Geosci Can 3:295–298

Vives i Batlle J, Aoyama M, Bradshaw C, Brown J, Buesseler KO, Casacuberta N, Christl M, Duffa C, Impens NREN, Iosjpe M, Masqué P, Nishikawa J (2018) Marine radioecology after the Fukushima Dai-ichi nuclear accident: are we better positioned to understand the impact of radionuclides in marine ecosystems? Sci Total Environ 618:80–92. https://doi.org/10.1016/j.scitotenv.2017.11.005

Yadav N, Pandey S, Gupta A, Dudani P, Gupta S, Rangarajan K (2023) Data privacy in healthcare: in the era of artificial intelligence. Indian Dermatol Online J 14:788–792. https://doi.org/10.4103/idoj.idoj_543_23

Yue S, Yang X (2024) Establishing a mechanism for international cooperation for Fukushima nuclear-contaminated water monitoring. Chin J Popul Resour Environ 22:20–33. https://doi.org/10.1016/j.cjpre.2024.03.003

Zalewska T (2014) Bioaccumulation of gamma emitting radionuclides in Polysiphonia fucoides. J Radioanal Nucl Chem 299:1489–1497. https://doi.org/10.1007/s10967-013-2794-2

Chapter 5
From Contaminants to Indicators: The Dual Function of Radionuclides in Marine Ecosystems

Karima Azoulay, Imane Bencheikh, Khadija El-Moustaqim, Najat Qisse, Dalal Boudoudou, Salah Eddine Sbai, and Jamal Mabrouki

Abstract The oceans, essential for climate regulation, biodiversity, and human resources, are under increasing pressure from anthropogenic pollution. This pollution comes primarily from urban, industrial, and agricultural activities, in addition to atmospheric and marine inputs such as oil spills and plastic waste. Pollutants include heavy metals, hydrocarbons, excess nutrients, and radionuclides, which compromise water quality, disrupt food chains, and threaten the food security of coastal populations. Transport pathways runoff, atmospheric deposition, ocean circulation, and sedimentation favor their accumulation in particularly vulnerable areas such as the Mediterranean.

Radionuclides occupy a unique position: as persistent contaminants, they are also powerful tracers for studying pollution dynamics and dating sedimentary deposits. Their fate depends on intrinsic properties (half-life, chemical affinity, decay mode) and environmental conditions (pH, salinity, suspended solids). Certain isotopes, such as ^{137}Cs or ^{210}Pb, can identify the origin of sediments, while tritium and carbon-14 are integrated into major biogeochemical cycles. In addition, bioaccumulating organisms mussels, fish, seagrass beds play a role as biological sentinels, revealing contamination and its effects on ecosystems.

The integration of radionuclides and bioindicators, combined with modern biomonitoring techniques, strengthens our understanding of transfer processes and provides strategic tools for managing and preventing environmental risks. This study places these approaches within the framework of biomonitoring, emphasizing the

K. Azoulay (✉) · I. Bencheikh · N. Qisse · D. Boudoudou · S. E. Sbai · J. Mabrouki
Laboratory of Spectroscopy, Molecular Modeling, Materials, Nanomaterial, Water and Environment, CERNE2D, Faculty of Science, Mohammed V University in Rabat, Rabat, Morocco
e-mail: karima.azoulay@um5r.ac.ma

K. El-Moustaqim
Improvement and Valuation of Plant Resources, Faculty of Sciences, Ibn Tofaïl University, Kenitra, Morocco

N. Mejjad (ed.), *Nuclear-Based Approaches application for marine pollution assessment and monitoring*, Environmental Science and Engineering,
https://doi.org/10.1007/978-3-032-12197-4_5

methodological integration to characterize ecological quality and guide environmental management.

Keywords Radionuclides · Marine pollutants · Bioindicators

5.1 Introduction

Oceans, essential for biodiversity, human resources and climate, are under increasing pressure from anthropogenic pollution (Daranagama and Liyanage 2024). Examples include heavy metals, hydrocarbons, excess fertilizers and radioactive substances, which pose direct threats to marine ecosystems and indirect risks to human health, particularly in coastal communities dependent on seafood (Daranagama and Liyanage 2024). Approximately nine million premature deaths per year are caused by the accumulation of these pollutants in the food chain, leading to serious health consequences, including causative diseases (Daranagama and Liyanage 2024).The increasing complexities of marine pollution therefore require a thorough understanding of the sources, behaviors and impacts of these contaminants, in order to develop effective prevention and management strategies (Fara and Muslim 2024). For my industrial discharges, they introduce heavy metals and persistent organic pollutants (POPs) into coastal areas, which accumulate in sediments, disrupt aquatic fauna and compromise food security (Nies 2018).Also, agricultural inputs, linked to the excessive use of fertilizers, cause nutrient pollution in waterways causing eutrophication and the proliferation of harmful algae which consequently deplete oxygen and destabilize marine ecosystems (Wijaya et al. 2024). In addition to radioactive pollution, which is a source of radionuclides from nuclear facilities. By accumulating in marine organisms, they constitute persistent risks for biodiversity and human health (Nies 2018). The consequences of radioactive contamination in continental and marine aquatic environments are profound and multiple. Radionuclides introduced by human activities, whether from atmospheric fallout, industrial discharges, or nuclear accidents, accumulate in soils, sediments, and water bodies, then circulate within ecosystems. This persistent presence disrupts the natural functioning of watercourses and increases the vulnerability of coastal areas, particularly those like the Mediterranean where the renewal of water bodies is limited. The fixation of isotopes such as cesium or plutonium on suspended particles alters the dynamics of sediment flows and leads to lasting contamination of alluvial margins and the seabed. Other radionuclides, such as tritium, are directly integrated into the major biogeochemical cycles of water and carbon, paving the way for more diffuse and sometimes insidious dissemination. These phenomena weaken the ecological stability of aquatic systems, reduce their self-purification capacity and increase the risks to biodiversity and human uses, particularly the quality of drinking water. Thus, radioactive pollution is not limited to a one-off threat, but represents a structural factor that determines the resilience and sustainability of ecosystems. Despite the urgency, sustainable understanding of marine radionuclides remains incomplete. Although recognized, systematic applications of radionuclides remain limited. This restricted approach compromises the development of robust marine models. Strengthening integrated approaches remains essential for sustainable monitoring. This chapter therefore explores the use of radionuclides as marine bioindicators. It analyzes their properties, sources of pollution and ecological impacts. Finally, it proposes to integrate these data into environmental frameworks.

5.2 From Contaminants to Indicators

5.2.1 Marine Pollution: Sources, Pathways, and Impacts

Marine pollution is currently a major environmental problem, largely dominated by land-based inputs (nearly 80%). The main sources include urban (untreated domestic wastewater), industrial (chemical effluents, heavy metals, PCBs, hydrocarbons), agricultural (excess nitrates and pesticides washed away by rainfall), and atmospheric deposition transporting persistent pollutants over long distances. In addition, there are marine sources, often spectacular but quantitatively less significant, such as oil spills, deliberate or accidental discharges from ships (degassing, ballast water), and, above all, floating or beached plastic waste (Goeury and Raymond 2014).

Transfer pathways to the marine environment are numerous: runoff and drainage through watersheds, atmospheric deposition carried by winds, transport and dispersion by ocean currents, and sedimentation and resuspension of pollutants. These mechanisms explain the concentration of waste in certain enclosed or semi-enclosed ocean areas, such as gyres, giving rise to "plastic continents." Ballast water and biofouling from ship hulls also play a crucial role in the spread of invasive species, disrupting local ecological balances (Goeury and Raymond 2014).

The impacts of this pollution are multiple and serious. Ecologically, we observe the proliferation of algae (green and red tides) and the formation of dead zones linked to eutrophication and hypoxia. Plastics, by fragmenting into micro- and nanoparticles, enter the food chain via plankton and concentrate persistent pollutants, thus contaminating all marine biodiversity. In terms of health, this contamination has repercussions on human populations, particularly coastal communities dependent on fishing and aquaculture. On the socio-economic level, the consequences include the collapse of fisheries, high costs of cleaning beaches and ports, and major losses for coastal tourism, compounded by image damage. Finally, marine pollution also compromises essential ecosystem services (climate regulation, carbon storage, fisheries productivity), amplifying climate disruption and threatening the long-term sustainability of the oceans (Goeury and Raymond 2014).

5.2.2 Pollutants in the Marine Environment

The marine environment is a prime receptacle for a wide variety of pollutants, whether of natural origin or resulting from human activities. Artificial radionuclides, introduced by atmospheric fallout from nuclear tests, major accidents such as Chernobyl or Fukushima, as well as regulated discharges from power plants and reprocessing sites, are among the most closely monitored contaminants. Once transferred by rivers, particularly the Rhône, these substances reach coastal areas where they interact with sediments, suspended particles, and living organisms. Their behavior depends heavily on their physicochemical properties: some, such as cesium 137, attach themselves to fine particles and follow sedimentary flows, while others, such as tritium, circulate mainly in dissolved form and become part of the water cycle. Marine pollutants also include heavy metals, hydrocarbons, and persistent organic compounds, which can act in synergy with radioactivity to affect ecosystems. The Mediterranean, due to its low water mass renewal and high vulnerability to continental inputs, is a particular example of this accumulation. The study of these pollutants not only allows us to assess the environmental impact of human activities, but also to use certain radionuclides as tracers to understand the dynamics of material flows and long-term contamination processes. In addition, hydrocarbons from oil spills alter water and sediments, causing disturbances for fish, invertebrates and plants, and impacting the balance of marine ecosystems (Rajak et al. 2024). Also, excess nitrogen and phosphorus cause eutrophication, depleting oxygen and altering marine ecosystems (Li et al. 2022).

5.2.3 Bioindicators in Marine Ecosystems

Bioindicators play a key role in assessing the health of marine ecosystems because they reflect, through their presence, concentration, or biological response, the impact of pollutants on the environment. Among them, certain radionuclides stand out as preferred tracers: ^{210}Pb, ^{137}Cs, and plutonium isotopes, thanks to their half-lives and differentiated behavior between dissolved and particulate phases, make it possible to reconstruct the origin of sediments, date recent deposits, and monitor the evolution of radioactive contamination. At the same time, living organisms such as mussels, fish, and seagrass beds constitute true "biological sentinels," capable of accumulating contaminants present in trace amounts in water or sediments. Their analysis provides information not only on the quality of the environment but also on potential risks to the food chain and, ultimately, to human health. These bioindicators therefore combine two major functions: that of scientific tools allowing a better understanding of the processes of transfer and storage of pollution, and that of operational indicators useful for environmental management and regulation (see (Eyrolle, EXTREMA contrat ANR-06-VULN-005 2011)). Thus, the integration of radionuclides and marine organisms as bioindicators strengthens our capacity to

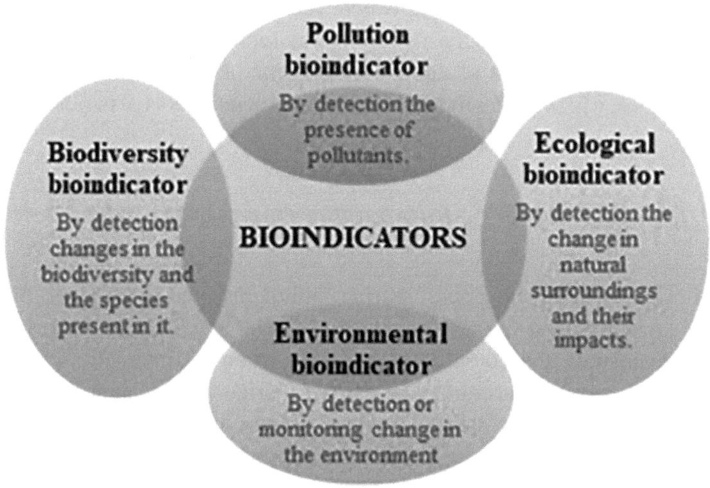

Fig. 5.1 Types of bioindicators for environmental assessment

monitor contamination, anticipate its effects and develop appropriate protection strategies (Fig. 5.1) (Eyrolle, EXTREMA contrat ANR-06-VULN-005 2011).

5.2.4 Radionuclides: Properties and Relevance

Radionuclides are essential for marine environmental studies. More specifically, in the study of tracers applied to understand pollution dynamics and ecosystem processes. Radionuclides can be of natural origin, such as uranium-238 and potassium-40, or anthropogenic, linked to energy production activities (Huang 2024). Cesium-137 and tritium are effective tracers for assessing and monitoring marine contamination (Randhawa 2023). The variety of radionuclide half-lives sheds light on ocean dynamics and pollutant dispersion (Li et al. 2022; Bograd et al. 2024).

Radionuclides are distinguished by fundamental properties that condition their behavior in the environment and explain their central role in water sciences and radioecology. Whether they are of natural origin resulting from the decay of uranium, thorium or potassium—or artificial, introduced by human activities such as nuclear tests, accidents (Chernobyl, Fukushima) or regulated discharges from installations, their presence in aquatic environments responds to different logics. Their fate depends both on intrinsic characteristics (radioactive half-life, decay mode, chemical affinity) and external factors such as pH, salinity, suspended load or the mineralogical nature of the soil. Some isotopes, such as ^{210}Pb or ^{137}Cs, constitute true “tracers” because their origins and decay dynamics make it possible to date sedimentary deposits or to follow the imprint of floods and solid flows. Others, such as tritium or carbon 14, are integrated into major biogeochemical cycles, providing unique insights into the exchanges between water, organic matter and living things. Thus, beyond their radioactivity, radionuclides represent powerful indicators for understanding the evolution of aquatic ecosystems and assessing the impacts of human activities, which makes them scientific and societal tools of major importance.

5.2.5 *Biomonitoring Techniques for Oceanic Health*

Biomonitoring techniques assess marine pollutants via multi-level biological responses (Antinero and Capangpangan 2024; Lehtonen 2024). Biomarkers measure toxic exposures and complex pollutant-environment interactions (He 2024). Radionuclides, isotopic tracers that assess the marine contaminant flux, allow the analysis of interactions between microplastics and contaminants. The latter often facilitate the migration and concentration of radionuclides (Ioannidis et al. 2023; Cresswell et al. 2020). In parallel, alternative bioindicators, particularly microbial or chemical, offer more accessible and less risky approaches. They make it possible to assess the impacts of microplastics on microbial communities or the increased presence of associated pollutants (Rosati et al. 2024). The joint use of radionuclides and bioindicators diversifies ecological assessment and strengthens environmental policies (Cresswell et al. 2020)

5.2.6 *Radionuclides as Strategic Indicators for Environmental Management and Risk Prevention*

Radionuclides should not be considered solely as contaminants, but also as genuine strategic indicators for pollution management and environmental risk prevention. Their presence and behavior in aquatic environments shed light on the dynamics of pollutants and provide a better understanding of transfer processes between soils, water, and sediments. Some isotopes serve as reliable temporal markers for reconstructing the history of contamination, while others reveal the current circulation of substances in ecosystems (Huang et al. 2025). Thanks to this tracing capacity, radionuclides provide essential information for assessing water quality, anticipating the impacts of extreme events such as floods, and measuring the consequences of industrial discharges on sensitive areas, particularly coastal areas. Their use in environmental monitoring programs illustrates the link between scientific research and public decision-making: they allow both a better understanding of natural mechanisms and provide tools to support management and regulation. In this sense, radionuclides present themselves as valuable allies in guiding policies for the protection of aquatic environments and strengthening resilience in the face of pollution risks (Eddy et al. 2025).

5.3 Conclusion

Marine pollution today represents a major challenge for the sustainability of the oceans and the human societies that depend on them. Radionuclides, although associated with environmental risks, also constitute valuable allies in understanding and managing this pollution. Their role as tracers, combined with the use of bioindicators and biomarkers, allows for better characterization of sources, transfer pathways, and ecological impacts. These integrated approaches strengthen monitoring capacity, inform public decisions, and support the implementation of effective environmental policies. Preserving the health of the oceans therefore requires not only limiting pollutant inputs, but also promoting radionuclides and bioindicators as strategic tools for prevention and environmental governance.

References

Antinero A, Capangpangan R (2024) Biomarkers: their vital role in understanding the complexity of marine pollution monitoring. J Environ Aquat Resour. https://doi.org/10.48031/msunjear.2024.07.03/

Bograd SJ, Anderson LC, Canonico G, Chiba S, Di Lorenzo E, Enterline C, Gorecki E, Griffis RB, Kleisner KM, Lachance H, Leinen M, Mills KE, Müller-Karger F, Roskar G, Schmidt J, Seary R, Seeyave S, Hwai TS, Soares J, Tigchelaar M (2024) Advancing the climate-biodiversity-fisheries nexus in the UN Decade of Ocean Science for Sustainable Development. ICES J Mar Sci. https://doi.org/10.1093/icesjms/fsae111

Cresswell T, Metian M, Fisher NS, Charmasson S, Hansman RL, Bam W, Bam W, Bock C, Swarzenski PW (2020) Exploring new frontiers in marine radioisotope tracing—adapting to new opportunities and challenges. Front Mar Sci 7:406. https://doi.org/10.3389/FMARS.2020.00406

Daranagama DA, Liyanage D (2024) Recent trends in microplastic detection based on machine learning and artificial intelligence. In: Fate of microplastics in wastewater treatment plants. CRC Press, pp 172–186. https://doi.org/10.1201/9781003438793-13

Eddy NO, Igwe O, Eze IS, Garg R, Akpomie K, Timothy C, Udeokpote G, Ucheana I, Paktin H (2025) Environmental and public health risk management, remediation and rehabilitation options for impacts of radionuclide mining. Discov Sustain 6:209. https://doi.org/10.1007/s43621-025-01047-6

Eyrolle, EXTREMA contrat ANR-06-VULN-005 (2011) Episodes météo climatiques extrêmes et redistribution des masses sédimentaires et des polluants associés au sein d'un système côtier. Rapport Final, Rapport DEI/SESURE 33

Fara M, Muslim M (2024) Application of nuclear analysis for bioaccumulation of microplastics with iodine-131 in marine organisms. World J Adv Eng Technol Sci 12(1):446–453. https://doi.org/10.30574/wjaets.2024.12.1.0251

Goeury D (2014) La pollution marine. In: Raymond W (ed) Mers et océans. Atlande, Clefs Concours, Paris

He Z (2024) Exploring the dangers of marine pollution to marine life. Highlights Sci Eng Technol 83:372–377. https://doi.org/10.54097/sn0d0n29

Huang J (2024) Research progress on sources, harms and countermeasures of marine pollution. E3S Web Conf 553:03005. https://doi.org/10.1051/e3sconf/202455303005

Huang W-H, Chen Z-M, Chen T-C, Yeh Y-L (2025) Assessing radiological risks of natural radionuclides on sustainable campus environment. Sustainability 17:691. https://doi.org/10.3390/su17020691

Ioannidis I, Kinigopoulou V, Giannakoudakis DA, Arkas M, Anastopoulos I, Triantafyllidis KS, Pashalidis I (2023) Microplastics and disposable face masks as "Trojan Horse" for radionuclides pollution in water bodies—a review with emphasis on the involved interactions. Sustain Chem Environ 1:100005. https://doi.org/10.1016/j.scenv.2023.100005

Lehtonen KK (2024) The use of biomarkers in monitoring and assessment of chemical contamination in the Baltic Sea. Proceedings. https://doi.org/10.3390/proceedings2023092080

Li Z, Chen R-C, Liu C, Xue Q, Wang Z, Zhou T (2022) Research on radionuclide diffusion mechanism in the ocean and emergency response under oceanic radioactive events. Sci Technol Nucl Install 2022:1–10. https://doi.org/10.1155/2022/6365560

Nies H (2018) Origin and management of radioactive substances in the marine environment. In: Salomon M, Markus T (eds) Handbook on marine environment protection. Springer, Cham, pp 733–753. https://doi.org/10.1007/978-3-319-60156-4_38

Rajak P, Ganguly A, Nanda S, Mandi M, Ghanty S, Das K, Biswas G, Sarkar S (2024) Toxic contaminants and their impacts on aquatic ecology and habitats. In: Spatial modeling of environmental pollution and ecological risk. Woodhead Publishing, pp 255–273. https://doi.org/10.1016/b978-0-323-95282-8.00040-7

Randhawa J (2023) Advanced analytical techniques for microplastics in the environment: a review. Bull Natl Res Cent 47:1–14. https://doi.org/10.1186/s42269-023-01148-0

Rosati L, Carraturo F, Capozzi F, Chianese T, La Pietra A, Salamone M, Spagnuolo V, Ferrandino I, Giordano S (2024) Microplastics' impact on the environment and the challenging selection of reliable key biomonitors. Water. https://doi.org/10.3390/w16182637

Wijaya H, Dien HA, Montolalu RI, Makapedua DM (2024) The dangers of marine pollution on consumer health, economic value, and aesthetic value in human life. Jurnal Ilmiah Platax (Edisi Elektronik) 12(2):103–109. https://doi.org/10.35800/jip.v12i2.55871

Chapter 6
Natural Radioactivity Concentrations Assessment of Sediment Cores of the Moulay Bousselham Coastal Lagoon, Morocco

Nezha Mejjad, Abdelmourhit Laissaoui, and Azzouz Benkdad

Abstract The present study aims to assess natural radioactivity in sediment cores collected from the Moulay Bousselham Lagoon. For this purpose, sediment cores were analyzed for radionuclides concentrations measurement by gamma spectrometry. At the same time, the radiometric dating approach was applied to establish the deposition history of radionuclides in the lagoon using Pb-210. In the three cores, Ra-226 levels show slight variations in the top layers and then deplete with depth. Similarly, for ^{232}Th and ^{40}K, the concentration activity decreases with depth, with a pronounced peak shown in the period extended between 1973 and 1980 in the three cores for the three studied radionuclides. This suggests a change in conditions controlling the accumulation of these naturally occurring radionuclides. The calculated radiation hazards index values were lower than the global average, signifying no radiological risk associated with the investigated radionuclides, and the lagoonal sediments do not present any serious hazard. This study is the first detailed database carried out in Moulay Bousselham on the assessment of the history of natural radionuclides deposition and concentrations, and it will be a reference for the activity of radionuclides in Moulay Bousselham lagoon sediment.

Keywords Sediment · Radionuclides · Grain size · Radiological hazards assessment · Moulay Bousselham lagoon

N. Mejjad (✉) · A. Laissaoui · A. Benkdad
National Centre for Nuclear Energy, Science and Technology (CNESTEN), Rabat, Morocco

N. Mejjad (ed.), *Nuclear-Based Approaches application for marine pollution assessment and monitoring*, Environmental Science and Engineering,
https://doi.org/10.1007/978-3-032-12197-4_6

6.1 Introduction

Coastal lagoons play an essential role as habitats and nurseries for multiple marine organisms and waterbirds. These ecosystems support different types of habitats, such as sandy beaches, intertidal mudflats, seagrass, mangroves, and salt marshes (Ayyam et al. 2019; Hobohm et al. 2021). They are very productive ecosystems where they provide many ecosystem services (Mejjad et al. 2023, 2025; Mateos-Molina et al. 2023; El Mahrad et al. 2020). Additionally, coastal lagoons are known as tourist destinations attracting economic investments (El Mahrad et al. 2022). Thus, different activities are practiced around these ecosystems, which consequently can influence the ecological functioning of these systems.

Lagoons, by their position at the land-ocean interface, are vulnerable to multiple pressures affecting their ecological functions (Lacoste et al. 2023; Celis-Hernandez et al. 2023; Pérez-Ruzafa et al. 2011, 2018). Numerous research studies and monitoring programs have been carried out to assess the health status of lagoon ecosystems to enhance their ecological and chemical status, such as the Beaufort Lagoon Ecosystems Long Term Ecological Research (BLE LTER) (https://ble.lternet.edu/) and the Marchica Observatory established to depollute and protect Marchica lagoon in Morocco (https://medtrix.fr/en/portfolio_page/nador-2/).

The human footprint in lagoon ecosystems is significant and diverse. Thus, multiple human activities affect coastal lagoons' environmental quality, including tourism, fishing, agricultural practices, and domestic and industrial activities. Lagoons can receive diverse types of pollutants from land-based sources including microplastics, heavy metals, persistent organic pollutants (POPs) and radioactive materials (Dalvand and Hamidian 2023; Shahroudi and Pourimani 2023; Specchiulli et al. 2023; Laissaoui et al. 2018; Benmhammed et al. 2021; Vargas-González et al. 2016; Mejjad et al. 2016; Guerzoni et al. 2007; Tsabaris et al. 2007).

Radionuclides are among the contaminants that occur naturally in plants, air, and soil (Ali et al. 2024; Lopez-Fernandez et al. 2021; Alharbi and El-Taher 2013; Zhu and Shaw 2000). Radionuclides can have natural and anthropogenic sources. Human activities, including industrial discharges, nuclear waste discharges, and nuclear accidents, have significantly contributed to radioactivity in coastal and marine environments (Yadav et al. 2025; Uchechukwu et al. 2025; Lloyd and Macaskie 2000). Radionuclides in the marine environment, similarly to other contaminants, are stored in the marine sediments. This latter is considered as a geochemical well of pollutants and widely used for pollution levels monitoring. Sediments are a valuable tool for assessing contamination in marine and coastal systems, and they are used to establish pollution history using radiometric dating.

The present chapter delivers data on natural radionuclides activity concentrations in sediments collected from Moulay Bousselham lagoon over time. To our knowledge, no previous studies have analyzed the radionuclides concentrations in this lagoon ecosystem. Thus, the present study is the first analysis of natural radioactive element concentrations in the Moulay Bousselham lagoon sediments. This study also uses radiological hazards index to assess the radiation exposure levels from naturally occurring radionuclides ^{226}Ra, ^{232}Th and ^{40}K, and thus provide quantitative data and measure of the possible health risks related to these exposures.

6.2 Methods

6.2.1 Study Area description

The Moulay Bousselham (MYB) Lagoon is located in the Northern part of the Moroccan Atlantic Coast and covers an area of 30 km^2. The lagoon comprises two Merjas, Merja Zerga (MZ) and Merja Kahla (MK). MK covers approximately 13 km^2 and is characterized by shallow waters, while at high tide, its depth does not exceed 50 cm (Fouad et al. 2019; Bazaïri et al. 2003). MZ covers around 27 km^2 with a higher depth (Bazaïri et al. 2003). It is known for its ecological and natural interest as a stop-over and wintering zone for various migratory birds. Also, the lagoon provides many ecosystem services, including fishing, coastal tourism, and climate change mitigation through the blue carbon ecosystem (seagrass: *Zostera noltei*) (Benmokhtar et al. 2023; Elso et al. 2017). Diverse activities, including tourism, fishing, and agriculture, are practiced along the lagoon. MYB Lagoon has been widely studied for heavy metal pollution assessment (El Idrissi et al. 2024; Anhichem et al. 2021; Meissara et al. 2014; Wariaghli et al. 2013; Maanan et al. 2013; Alaoui Mhamdi et al. 2010; Hmama et al. 2001), where they highlighted the effects of human activities development since the 1970s on this ecosystem, including water quality, ecology, and biodiversity.

6.2.2 Sampling Procedures

Three sediment cores were collected from Moulay Bousselham lagoon using a PVC tube or PVC corer of a diameter of 5 cm and a length of 60 cm. The first sediment core was collected from Merja Kahla (C1MK), while C2MZ and C3MZ were retrieved from Merja Zerga (Fig. 6.1). Sampling points were selected based on

Fig. 6.1 Moulay Bousselham Lagoon's geographical location and the sediment cores' sampling point

sediment distribution and hydrodynamic parameters (Maanan et al. 2013). Sediments were sectioned into layers of 2 cm, dried, ground, and sieved for further analysis.

6.2.3 Physico-Chemical Parameters Determination

Organic matter content (OM) was determined for each layer of sediment. Loss on ignition (LOI) at 500 °C method was used to measure OM %. The method is based on igniting 5 g of the sediment sample in a muffle furnace at a temperature of 500 °C for 2 h. The percentage of OM is calculated using the following equation (Eq. 6.1):

$$\boldsymbol{OM}(\%) = \frac{\text{Dry mass of the sample before ignition} \\ -\text{Dry mass of the sample after ignition}}{\text{Dry mass of the sample before ignition}} * 100 \quad (6.1)$$

The Mastersizer 2000 laser diffraction particle size analyzer determined the grain size distribution. The grain size distribution analysis was performed for each sediment core sample using Gradistat, a computer program (Blott and Pye 2001). Statistical parameters such as mean, skewness, sorting, and kurtosis were determined using this computer program following the Folk and Ward method (Folk and Ward 1957).

6.2.4 Radionuclide Determination Using Gamma Spectrometry

The concentration activity of gamma-emitting radioelement in each sediment sample was measured using gamma spectrometry with a high-purity Germanium detection system of a relative efficiency of 30%. The detector's energy resolution is 1.8 keV at an energy of 1332.5 keV. A standard was prepared for calibration purposes. This standard calibrates the system measurement and presents a counting geometry similar to that of the samples under investigation. ^{226}Ra, ^{232}Th, and ^{40}K concentrations were determined for each layer of sediment cores. The specific activity of ^{226}Ra was measured using the photopeaks of ^{214}Pb (295.2 and 351.9 keV) and of ^{214}Bi (609.3 keV), while for ^{40}K, the activity level was determined at 1460.8 keV. To determine ^{232}Th activity concentration, ^{228}Ac (911.2 keV) photopeak was used.

Using $^{210}Pb_{ex}$ and ^{137}Cs, the three sediment cores were dated. The Constant Rate of Supply (CRS) model was applied to profiles of ^{210}Pbex to calculate the age of each layer of the sediment cores from the total inventory (in Bq m^{-2}). This later is calculated as the sum of the product along the sediment core of activity, bulk density, and thickness of each layer.

6.2.5 Radiological Hazards Assessment

Multiple indices were used to assess the radiological hazards in the samples of three sediment cores. These indices include radium equivalent activity (Ra_{eq}), the total absorbed gamma dose rate (D), external (H_{ex}), and internal hazard (H_{in}) indices.

- **(D) is the total absorbed gamma dose rate** (nGy h^{-1}) in outdoor air at 1 m above the ground. D is calculated using the following equation (Eq. 6.2) (United Nations Scientific Committee on the Effects of Atomic Radiation 2000):

$$D = 0.462 \times A_{Ra} + 0.604 \times A_{th} + 0.0417 \times A_K \quad (6.2)$$

- where A_{Ra}, A_{Th}, A_K are the activity concentrations in Bq kg^{-1} of ^{226}Ra, ^{232}Th and ^{40}K respectively.
- **The annual effective dose rate E (mSv year^{-1})** received by the population is estimated using the equation below (Eq. 6.3) (UNSCEAR, 2000):

$$E\left(mSv\,year^{-1}\right) = \mathrm{D}\left(\mathrm{nGy}\,h^{-1}\right) \times 8760\left(h\,year^{-1}\right) \times 0.2 \times 0.7\left(\mathrm{Sv\,Gy}^{-1}\right) \times 10^{-6} \quad (6.3)$$

- where D (nGy h^{-1}) is the absorbed dose rate in air, 8760 h means the time or number of hours in 1 year, 0.7 (Sv Gy^{-1}) is the conversion factor, it converts the absorbed dose rate in air to human effective dose. 0.2 is the outdoor occupancy factor (United Nations Scientific Committee on the Effects of Atomic Radiation 2000).
- **The Radium Equivalent Activity (Ra_{eq})** is used to compare the activities of sediment samples comprising different activities and is calculated using Eq. (6.4) (Boukhenfouf and Boucenna 2011; Alshahri and Alqahtani 2015):

$$Ra_{eq} = A_{Ra} + 1.43 \times A_{Th} + 0.077 \times A_K \quad (6.4)$$

- This relation assumes that 370 Bq kg^{-1} of ^{226}Ra, 259 Bq kg^{-1} of ^{232}Th, and 4810 Bq kg^{-1} of ^{40}K each independently produces the same gamma dose rate.
- To analyze the significance of external radiation exposure from studied sediments, the **external hazard index (H_{ex})** is used. Hex is calculated using the following equation (Eq. 6.5) (Zare et al. 2012; AlZahrani et al. 2011):

$$H_{ex} = A_{Ra} \div 370 + A_{Th} \div 259 + A_K \div 4810 \quad (6.5)$$

- **Internal hazard index (H_{in})** is used to estimate the internal exposure to Radon. It is calculated using the equation below (Eq. 6.6) (Beretka and Mathew 1985):

$$H_{in} = A_{Ra} \div 185 + A_{Th} \div 259 + A_K \div 4810 \quad (6.6)$$

- H_{ex} and H_{in} must be less than unit to be considered acceptable or have negligible hazardous effects on the population.

6.3 Results and Discussions

6.3.1 Descriptive Statistics of Sedimentological Parameters

6.3.1.1 Organic Matter Content

Table 6.1 shows descriptive statistics of organic matter (OM) and grain size in three sediment cores. OM content shows an important value in the sediment core of Merja Kahla (C1MK) compared with sediment cores collected from Merja Zerga (C2MZ and C3MZ) with mean values of 5.32%, 4.26%, and 2.95%, respectively. The Maximum value (7.9%) was recorded in C1MK while the lowest value is registered in C3MK (4.32%). These values are close to the OM percentages reported in previous studies (e.g., 0 and 10% in Boutoumit et al. (2021) and 1.01% and 10.59% in summer in Anhichem et al. (2021)). Different natural and anthropogenic factors can influence OM

Table 6.1 Descriptive statistics of organic matter content (OM) and grain size (Mud and Sand)

	N	Mean	SD	Sum	Min	Max	Texture	Sorting
Core 1 Merja Kahla								
OM (%)	24	**5.32**	1.12	127.88	3.85	**7.90**	–	–
SAND (%)	24	74.04	8.42	1777.06	57.97	**91.15**	Muddy sand	Very poorly sorted and poorly sorted
MUD (%)	24	25.95	8.42	622.94	8.84	42.02	Silty sand	
Core 2 Merja Zerga								
OM (%)	20	4.26	0.73	85.22	2.65	5.7	–	–
SAND (%)	20	60.4	23.09	1207.7	13.35	84.54	Muddy sand	Very poorly sorted and poorly sorted
MUD (%)	20	**39.61**	23.09	792.3	15.45	**86.64**	Silty sand	
Core 3 Merja Zerga								
OM (%)	26	2.95	0.71	76.71	1.7	4.32	–	–
SAND (%)	26	**76.15**	11.51	1979.9	51.4	**92.25**	Muddy sand	Very poorly sorted and poorly sorted
MUD (%)	26	23.84	11.51	620.09	7.74	48.6	Silty sand	

Bold values indicate high values or content

content in sediment, including terrestrial inputs and marine productivity (phytoplankton, algae, and other organisms). Natural factors such as grain size and hydrodynamic conditions can also affect OM content, where it is known that finer fractions have a tendency to accumulate OM more than coarser fractions (Yang et al. 2021). Noting that Merja Zerga receives terrestrial input from human activities installed around it, but it shows less important OM content, which may be explained by the hydrodynamic conditions, which are more important in MZ compared to MK characterized by low hydrodynamic conditions favoring OM-rich sediments deposition (Puppin et al. 2024; Lu et al. 2020).

6.3.1.2 Grain Size Distribution

The three sediment cores show similar textural groups, "muddy sand and silty sand," all along the cores. C2MZ recorded the maximal percentage of Mud amount (86.64%) in deeper layers of sediment core while the range values of sand content registered in both C1MK and C3MZ fluctuated between 51% and 92% all along the cores. The sediment of the MK core is composed of very coarse Silty Coarse Sand in the top layers (0–16 cm) and then very coarse Silty Medium Sand in the deeper layers (16–48 cm). Similarly, C2MZ and C3MZ show very coarse Silty Coarse Sand, very coarse Silty fine sand, and very coarse Silty medium sand fluctuating all along the cores, suggesting inhomogeneous grain size. Based on statistical parameters, the sediments of both Merjas are very poorly sorted and poorly sorted, signifying that these sediments contain different ranges of grain sizes (varied sizes), which are all mixed together (Lewis and McConchie 2012).

6.3.2 Activity Concentration of Radionuclides in Sediment Cores

The vertical distribution of activity concentration of ^{226}Ra, ^{232}Th, and ^{40}K is depicted in Fig. 6.2. In each core, the three studies' radionuclides indicate similar vertical distribution where the activities fluctuate in the top layers and then deplete with depth. The concentrations of ^{226}Ra are lower than 13 Bq kg^{-1} in C1MK and C2MZ, while they reach a maximal value of 22 Bq kg^{-1} in the sediment layer corresponding to the year 1973 in C3MK core. In the three cores, ^{226}Ra levels show some fluctuations in the top layers and then deplete with depth. Similarly, for ^{232}Th and ^{40}K, the concentration activity decreases with depth, with a pronounced peak shown in the period extended between 1973 and 1980 in the three cores for the three studied radionuclides. This suggests a change in conditions controlling the accumulation of these naturally occurring radionuclides. It is worth noting that the lagoon has known several changes within this period, with the damming of the Dradar River in 1979, the closing and opening of the lagoon, and the rapid development of urbanization and human activities since the 1970s (Maanan et al. 2013).

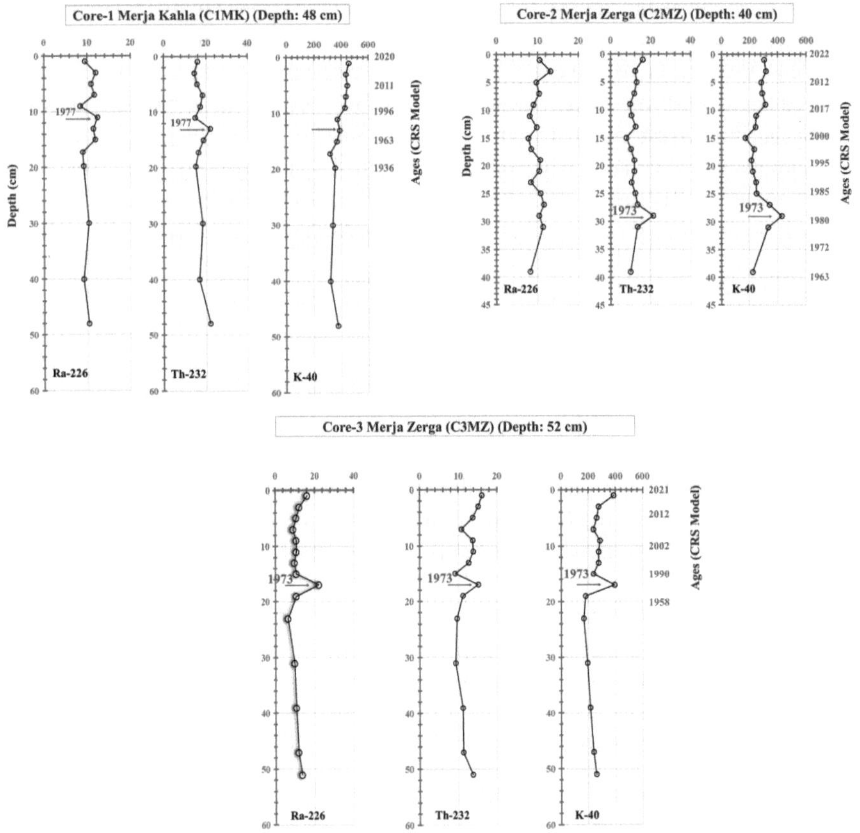

Fig. 6.2 The vertical distribution of concentrations activity of ^{226}Ra, ^{232}Th, and ^{40}K (Bq kg^{-1}) measured in each layer of the three sediment cores versus ages provided by the applied CRS model

However, the recorded values remain less important than those reported in a similar ecosystem worldwide (Table 6.2) and thus show values close to values found in coastal systems considered non-contaminated (Laissaoui et al. 2013). ^{226}Ra shows values less important than those recorded in Moroccan lagoons sediments such as Sidi Moussa lagoon (75 Bq kg^{-1}; (Benmhammed et al. 2021)) and Oualidia lagoon (30.20 Bq kg^{-1}; (Mejjad et al. 2016)). In three sediment cores, ^{226}Ra and ^{232}Th indicated similar values. At the same time, ^{40}K exhibited a high value in the core collected from Merja Kahla, exceeding values recorded in the cores from Merja Zerga, Makoko Lagoon (Nigeria), Nile River (Egypt) and recommended value by UNSCEAR (United Nations Scientific Committee on the Effects of Atomic Radiation 2000). This significant value recorded in the top layer of C3MK can have geochemical origins, where ^{40}K is known as an important radionuclide, representing 0.012% of naturally occurring potassium (Paschoa and Steinhäusler 2010). Also,

Table 6.2 Comparison of average activity concentrations of natural radionuclides in the sediment samples of Moulay Bousselham Lagoon with those around the world

Site	Ra-226	Th-232	K-40	References
MYB Lagoon (C1MK), Morocco (surface layer)	9.5	16	**456.7**	Present study
MYB Lagoon (C2MZ), Morocco (surface layer)	10.6	16.2	305.8	Present study
MYB Lagoon (C3MZ), Morocco (surface layer)	16.2	16.2	387.2	Present study
Saudi coastline of the Gulf of Aqaba	11.4	22.5	641.1	Al-Trabulsy et al. (2011)
Niger Delta (Biseni) flood plain lakes, Nigeria	20	20	180	Agbalagba and Onoja (2011)
İkizdere Valley, Turkey	–	32.71	811.68	Keser et al. (2013)
Tam Giang-Cau Hai, Thi Nai, and Nai lagoons in the Center of Vietnam	34.15	67.22	675.78	Nhon et al. (2024)
Makoko Lagoon Nigeria	10.15	4.39	41.04	Jibiri et al. (2023)
Nile River, Egypt	16.30	12.94	200.1	ElTaher et al. (2019)
Greece Aegean Sea	15	25	565	Shahrokhi et al. (2021)
Coast of Oman	20.49	2.26	44.83	(Al Shaaibi et al. 2021)
East Coast region, Cyprus	23	19	682.1	Abbasi et al. (2020)
Mediterranean Sea coast, North Cyprus	20.1	18.4	467.3	Abbasi and Mirekhtiary (2020)
UNSCEAR	33	45	420	United Nations Scientific Committee on the Effects of Atomic Radiation (2000)

Bold values indicate high values or content

the considerable concentration of ^{40}K can be related to terrestrial input, as coastal lagoons receive substantial inputs of continental eroded materials via surface runoff and rivers, which transport potassium-rich minerals to marine sediments (Paschoa and Steinhäusler 2010; Pandion et al. 2025). In contrast, the values of ^{40}K recorded in the studied sediment cores remain less important than those found in similar coastal systems such as in the Saudi coastline of the Gulf of Aqaba (Al-Trabulsy et al. 2011), in İkizdere Valley, Turkey (Keser et al. 2013) and in Tam Giang-Cau Hai, Thi Nai, and Nai lagoons in the Center of Vietnam (Nhon et al. 2024).

Noting that no industrial activities are located near the lagoon can cause enrichment by radionuclides, thus explaining the low obtained values. The recorded changes and fluctuations of the radionuclide activity concentrations were well correlated to the natural and anthropogenic changes in the lagoon, which suggests that these variations over time are related to various disturbances that occurred in the lagoon.

6.3.3 Radiological Hazard Parameters

6.3.3.1 $H_{(in)}$, $H_{(ex)}$

The internal (H_{in}) and external (H_{ex}) hazards index were calculated for each layer of three sediment cores and shown in Fig. 6.3 versus depth and ages. Both H_{in} and H_{ex} showed values less than unity all along the cores, with a pronounced peak corresponding to the periods between 1973 and 1980. The range values for H_{in} were 0.18–0.23, 0.11–0.23, and 0.11–0.26 for C1MK, C2MZ, and C3MZ, respectively. The calculated values for Hex ranged from 0.15 to 0.20, 0.09 to 0.20, and from 0.09 to 0.20 for C1MK, C2MZ, and C3MZ, respectively. All calculated values are below 1, indicating that there is an insignificant risk of radon and its short-lived progeny on the human respiratory organs because of the low values of gamma dose rate in all investigated sediment layers.

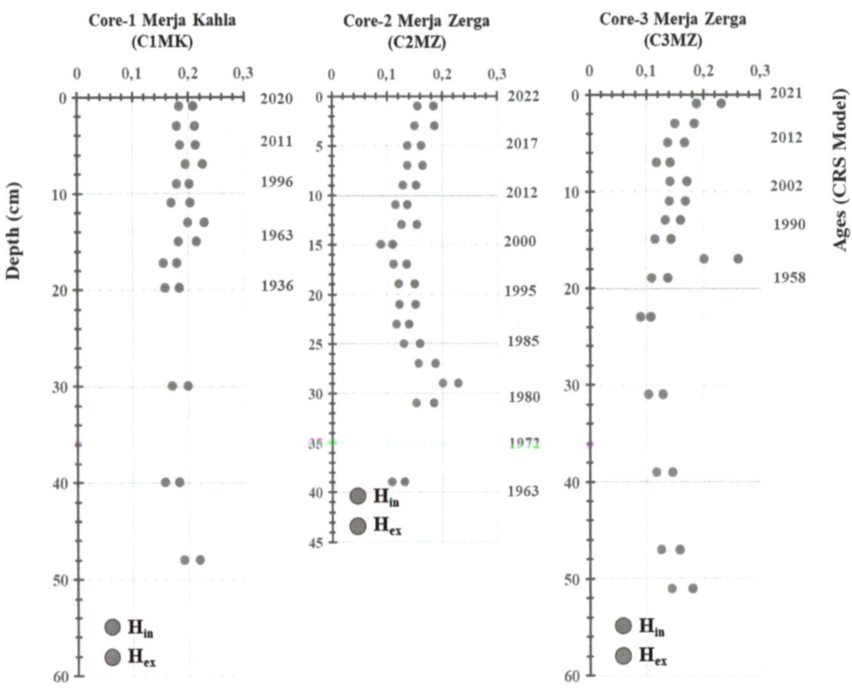

Fig. 6.3 Vertical distribution of H(in) and H(ex) indices versus depth and ages provided by the CRS Model (Values must be <1)

6.3.3.2 Radium Equivalent Activity (Ra_{eq}), Absorbed Gamma Radiation Dose Rate in the Air (D), and Annual Effective Dose (E)

Radium equivalent activity (Ra_{eq}), absorbed gamma radiation dose rate in the air (D), and annual effective dose (E) were calculated for each layer of the three cores. Table 6.3 shows the Raeq, D, and E indexes as mean values at different depths. The radium equivalent activity values fluctuate between 57.03 and 73.04 Bq kg^{-1}, 32.4 and 74.4 Bq kg^{-1}, 33.2 and 74.23 Bq kg^{-1} for C1MK, C2MZ, and C3MZ, respectively. The recorded values were much lower than the recommended value by UNSCEAR (1993) of 370 Bq kg^{-1}. The mean values of annual effective dose (E) and gamma absorbed dose rate (D) in air calculated at different depths of the cores were all considerably below the acceptable values defined by UNSCEAR (1993).

Table 6.3 Mean values of radium equivalent activity (Bq kg^{-1}), absorbed gamma radiation dose rate in air (nGy h^{-1}), and annual effective dose (mSv $year^{-1}$) at different depths

Depth (cm)	Raeq (Bq kg^{-1})	D (nGy h^{-1})	E (mSv y^{-1})	Ages (CRS Model)
Core 1 Merja Kahla				
0–10 cm	67.82	33.10	0.04	2020–1996
10–15 cm	67.55	32.48	0.04	1996–1963
15–20 cm	57.63	27.80	0.03	1963–1936
20–48 cm	63.89	30.57	0.04	–
Core 2 Merja Zerga				
0–10 cm	52.06	25.18	0.03	2022–2012
10–14 cm	44.51	21.42	0.03	2012–2004
14–16 cm	32.39	15.56	0.02	2000
15–20 cm	42.78	20.51	0.03	2000–1995
20–28 cm	48.32	23.28	0.03	1995–1982
28–30 cm	74.10	35.66	0.04	1982–1980
30–40 cm	48.26	23.34	0.03	1980–1963
Core 3 Merja Zerga				
0–10 cm	54.16	26.02	0.03	2021–2002
10–16 cm	47.73	22.99	0.03	2002–1983
16–18 cm	74.23	35.84	0.04	1973
18–22 cm	40.35	19.12	0.02	1973–1930
22–40 cm	38.13	18.22	0.02	–
40–51 cm	49.93	23.89	0.03	–
World's average (UNSCEAR 2000, 1993)	370	55	1	–

6.4 Conclusions

The present chapter provides the first data on natural radionuclides concentration distribution over time in the Moulay Bousselham Lagoon sediments. The activity concentrations of ^{226}Ra, ^{232}Th, and ^{40}K in 70 sediment samples were almost below the acceptable values recommended by UNSCEAR. However, ^{40}K in the surface layer of the sediment core C3MZ slightly exceeded the UNSCEAR recommended values for the uncontaminated area, which can be attributed to geochemical origins, including ^{40}K natural abundance and/or terrestrial inputs. The obtained radiation hazards index values were lower than the global average, signifying no radiological risk associated with the investigated radionuclides.

The findings can be used as reference data, mainly the study quantified naturally occurring radionuclides historical record by determining ^{226}Ra, ^{232}Th, and ^{40}K activity-concentrations in Moulay Bouselham lagoon sediments.

References

Abbasi A, Mirekhtiary F (2020) Heavy metals and natural radioactivity concentration in sediments of the Mediterranean Sea coast. Mar Pollut Bull 154:111041

Abbasi A, Zakaly HM, Mirekhtiary F (2020) Baseline levels of natural radionuclides concentration in sediments East coastline of North Cyprus. Mar Pollut Bull 161:111793

Agbalagba E, Onoja R (2011) Evaluation of natural radioactivity in soil, sediment and water samples of Niger Delta (Biseni) flood plain lakes, Nigeria. J Environ Radioact 102(7):667–671

Al Shaaibi M, Ali J, Duraman N, Tsikouras B, Masri Z (2021) Assessment of radioactivity concentration in intertidal sediments from coastal provinces in Oman and estimation of hazard and radiation indices. Mar Pollut Bull 168:112442

Alaoui Mhamdi A, Choura M, Maanan M, Zourarah B, Robin M, Conceição MF, Carruesco C (2010) Metal fluxes to the sediments of the Moulay Bousselham lagoon, Morocco. Environ Earth Sci 61:275–286

Alharbi A, El-Taher A (2013) A study on transfer factors of radionuclides from soil to plant. Life Sci J 10(2):532–539

Ali S, Baloch SB, Bernas J, Konvalina P, Onyebuchi EF, Naveed M, Ali H, Jamali ZH, Nezhad MTK, Mustafa A (2024) Phytotoxicity of radionuclides: a review of sources, impacts and remediation strategies. Environ Res 240:117479

Alshahri F, Alqahtani M (2015) Chemical fertilizers as a source of ^{238}U, ^{40}K, ^{226}Ra, ^{222}Rn and trace metal pollutant of the environment in Saudi Arabia. Environ Sci Pollut Res 22:8339–8348

Al-Trabulsy HA, Khater AEM, Habbani FI (2011) Radioactivity levels and radiological hazard indices at the Saudi coastline of the Gulf of Aqaba. Radiat Phys Chem 80(3):343–348

AlZahrani JH, Alharbi WR, Abbady AGE (2011) Radiological impacts of natural radioactivity and heat generation by radioactive decay of phosphorite deposits from northwestern Saudi Arabia. Aust J Basic Appl Sci 5(6):683–690

Anhichem M, Benbrahim S, Idrissi M, Chfiri R, Yahyaoui A (2021) Environmental state of the Merja Zerga lagoon in Moulay Bousselham, Morocco: metallic contamination levels and origins in sediments. Aquac Aquar Conserv Legis 14(2):774–785

Ayyam V, Palanivel S, Chandrakasan S (2019) Coastal ecosystems and services. In: Coastal ecosystems of the tropics—adaptive management. Springer, pp 21–47

Bazaïri H, Bayed A, Glémarec M, Hily C (2003) Spatial organisation of macrozoobenthic communities in response to environmental factors in a coastal lagoon of the NW African coast (Merja Zerga, Morocco). Oceanol Acta 26(5–6):457–471

Benmhammed A, Laissaoui A, Mejjad N, Ziad N, Chakir E, Benkdad A, El Yahyaoui A (2021) Recent pollution records in Sidi Moussa coastal lagoon (western Morocco) inferred from sediment radiometric dating. J Environ Radioact 227:106464

Benmokhtar S, Robin M, Maanan M, Boutoumit S, Badaoui B, Bazairi H (2023) Monitoring the spatial and interannual dynamic of Zostera noltei. Wetlands 43(5):43

Beretka J, Mathew PJ (1985) Natural radioactivity of Australian building materials, industrial wastes and by-products. Health Phys 48(1):87–95

Blott SJ, Pye K (2001) GRADISTAT: a grain size distribution and statistics package for the analysis of unconsolidated sediments. Earth Surf Process Landf 26(11):1237–1248

Boukhenfouf W, Boucenna A (2011) The radioactivity measurements in soils and fertilizers using gamma spectrometry technique. J Environ Radioact 102(4):336–339

Boutoumit S, Bououarour O, El Kamcha R, Pouzet P, Zourarah B, Benhoussa A, Bazairi H (2021) Spatial patterns of macrozoobenthos assemblages in a sentinel coastal lagoon: biodiversity and environmental drivers. J Mar Sci Eng 9(5):461

Celis-Hernandez O, Ávila E, Rendón-von Osten J, Briceño-Vera EA, Borges-Ramírez MM, Gómez-Ponce AM, Capparelli VM (2023) Environmental risk of microplastics in a Mexican coastal lagoon ecosystem: anthropogenic inputs and its possible human food risk. Sci Total Environ 879:163095

Dalvand M, Hamidian AH (2023) Occurrence and distribution of microplastics in wetlands. Sci Total Environ 862:160740

El Idrissi S, Labriki A, Sadiki M, Akhlidej N, Bounouira H, Said S, Chakiri S (2024) Geochemical analysis of coastal sediments near Larache and Moulay Bousselham, Morocco. Ecol Eng Environ Technol 25:221–231

El Mahrad B, Abalansa S, Newton A, Icely JD, Snoussi M, Kacimi I (2020) Social-environmental analysis for the management of coastal lagoons in North Africa. Front Environ Sci 8:37

El Mahrad B, Newton A, Murray N (2022) Coastal lagoons: important ecosystems. Front Young Minds 10:637578

Elso MZ, Manent P, Luque A, Ramdani M, Robaina RR (2017) Genetic description and remote sensing techniques as management tools for Zostera noltii seagrass populations along the Atlantic Moroccan coast. J Coast Res 33(1):78–87

ElTaher A, Najam L, Hussian I, Ali Omar MA (2019) Evaluation of natural radionuclide content in Nile River sediments and excess lifetime cancer risk associated with gamma radiation. Iran J Med Phys 16(1):27–33

Folk RL, Ward WC (1957) Brazos River bar [Texas]; a study in the significance of grain size parameters. J Sediment Res 27(1):3–26

Fouad M, Bessi H, Benhra A, Bouhallaoui M (2019) The benthic community used as bioindicator for assessment of the quality of a Moroccan coastal ecosystem: Moulay Bousselham Lagoon. Int J Eng Res Technol 8(3):127–133

Guerzoni S, Rossini P, Sarretta A, Raccanelli S, Ferrari G, Molinaroli E (2007) POPs in the Lagoon of Venice: budgets and pathways. Chemosphere 67(9):1776–1785

Hmama N, El Youssi M, Fekhaoui M (2001) Contamination métallique de la nappe aquifère de la lagune de Moulay Bouselham. Bull Inst Sci Rabat 23:89–93

Hobohm C, Schaminée J, van Rooijen N (2021) Coastal habitats, shallow seas and inland saline steppes: ecology, distribution, threats and challenges. In: Coastal habitats conference. Springer, Cham, pp 279–310

Jibiri NN, Ugbechie A, Sowunmi AA, Akomolafe IR (2023) Radionuclide contents in sediment and seafood from Makoko Lagoon, Lagos State, Nigeria. Mar Pollut Bull 192:114992

Keser RECEP, Görür FK, Alp İ, Okumusoglu NT (2013) Determination of radioactivity levels and hazards of sediment and rock samples in İkizdere and Kaptanpasa Valley, Turkey. Int J Radiat Res 11(3):155

Lacoste É, Jones A, Callier M et al (2023) A review of knowledge on the impacts of multiple anthropogenic pressures on the soft-bottom benthic ecosystem in Mediterranean coastal lagoons. Estuar Coasts 46:2190–2207
Laissaoui A, Mas JL, Hurtado S, Ziad N, Villa M, Benmansour M (2013) Radionuclide activities and metal concentrations in sediments of the Sebou Estuary, NW Morocco, following a flooding event. Environ Monit Assess 185:5019–5029
Laissaoui A, Mejjad N, Ziad N, Ait Bouh H, El Hammoumi O, Benkdad A, Fekri A (2018) Evidence for a recent increase in delivery of atmospheric ^{210}Pb to Oualidia lagoon, coastal Morocco. Environ Monit Assess 190(11):642
Lewis DW, McConchie D (2012) Analytical sedimentology.1st edn. Springer Science & Business Media
Lloyd JR, Macaskie LE (2000) Bioremediation of radionuclide-containing wastewaters. Environ Microbe-Metal Interact Am Soc Microbiol 77:277–327
Lopez-Fernandez M, Jroundi F, Ruiz-Fresneda MA, Merroun ML (2021) Microbial interaction with and tolerance of radionuclides: underlying mechanisms and biotechnological applications. Microb Biotechnol 14(3):810–828
Lu X, Zhou F, Chen F, Lao Q, Zhu Q, Meng Y, Chen C (2020) Spatial and seasonal variations of sedimentary organic matter in a subtropical bay: implication for human interventions. Int J Environ Res Public Health 17(4):1362
Maanan M, Landesman C, Maanan M, Zourarah B, Fattal P, Sahabi M (2013) Evaluation of the anthropogenic influx of metal and metalloid contaminants into the Moulay Bousselham lagoon, Morocco, using chemometric methods coupled to geographical information systems. Environ Sci Pollut Res 20:4729–4741
Mateos-Molina D, Pittman SJ, Antonopoulou M, Carpenter S, Möller M, Muzaffar SB, Bejarano I (2023) Coastal lagoons (Khors) of the emirates. In: A natural history of the emirates. Springer, Cham, pp 241–265
Meissara A, Fekhaoui M, El Morhit M, Waeles M, El Abidi A, Yahyaoui A (2014) Tissular distribution of heavy metals in the Mugil cephalus of the Moulay Bousselham lagoon (Morocco). Int J Sci Knowl 3(1):18–26
Mejjad N, Laissaoui A, El-Hammoumi O, Benmansour M, Benbrahim S, Bounouira H, Bounakhla M (2016) Sediment geochronology and geochemical behavior of major and rare earth elements in the Oualidia Lagoon in western Morocco. J Radioanal Nucl Chem 309:1133–1143
Mejjad N, Safhi AEM, Laissaoui A, El Aouidi S, Hilal I (2023) Valuing the ecological and socio-economic interests of the Oualidia Lagoon (Morocco): an ecosystem services framework. Eng Proc 56(1):77
Mejjad N, Safhi AEM, Laissaoui A (2025) Insightful analytical review of potential impacts of microplastic pollution on coastal and marine ecosystem services. J Hazard Mater Adv 17:100578
Nhon DH, Van Quan N, Hai PS, Van Vuong B, Anh NN, Ve ND, Chien HT (2024) Assessment of radioactivity and radiological risk indices in the sediments of the Tam Giang-Cau Hai, Thi Nai, and Nai lagoons in the Center of Vietnam. Radiochim Acta 112(9):679–689
Pandion K, Arunachalam KD, Gaafar ARZ, Almunqedhi BM, Kuppusamy S, Singh RP, Balasubramani R (2025) Assessment of naturally occurring radionuclides in the coastal sediment with statistical analysis of radiological risk parameters. J Hydroinformatics 25:17–32
Paschoa A, Steinhäusler F (2010) Terrestrial, atmospheric, and aquatic natural radioactivity. Radioactivity Environ 17:29–85
Pérez-Ruzafa A, Marcos C, Pérez-Ruzafa I (2011) Mediterranean coastal lagoons in an ecosystem and aquatic resources management context. Phys Chem Earth Parts A/B/C 36(5–6):160–166
Pérez-Ruzafa A, Pérez-Ruzafa IM, Newton A, Marcos C (2018) Coastal lagoons: environmental variability, ecosystem complexity, and goods and services uniformity. In: Coasts and estuaries. Elsevier, pp 253–276

Puppin A, Tognin D, Ghinassi M, Franceschinis E, Realdon N, Marani M, D'Alpaos A (2024) Spatial patterns of organic matter content in the surface soil of the salt marshes of the Venice Lagoon (Italy). Biogeosciences 21(12):2937–2954

Shahrokhi A, Adelikhah M, Chalupnik S, Kovács T (2021) Multivariate statistical approach on distribution of natural and anthropogenic radionuclides and associated radiation indices along the north-western coastline of Aegean Sea, Greece. Mar Pollut Bull 163:112009

Shahroudi SMM, Pourimani R (2023) Pollution distribution caused by natural and artificial radionuclide in water and sediments of the International Miankaleh Lagoon in north of Iran. Environ Adv 13:100413

Specchiulli A, Pastorino P, De Rinaldis G, Scirocco T, Anselmi S, Cilenti L, Ungaro N, Renzi M (2023) Multiple approach for assessing lagoon environmental status based on water bodies quality indices and microplastics accumulation. Sci Total Environ 892:164228

Tsabaris C, Eleftheriou G, Kapsimalis V, Anagnostou C, Vlastou R, Durmishi C, Kalfas CA (2007) Radioactivity levels of recent sediments in the Butrint Lagoon and the adjacent coast of Albania. Appl Radiat Isot 65(4):445–453

Uchechukwu NB, Obinna OE, Jindal MK, David A, Onomakere AG, Norbert JN, Fabian EI (2025) Impact of radioactive elements and oceanographic factors on marine ecosystems. In: Radiation status in the marine world. Springer, Cham, pp 193–226

United Nations Scientific Committee on the Effects of Atomic Radiation (2000) Sources and effects of ionizing radiation, United Nations Scientific Committee on the Effects of Atomic Radiation (UNSCEAR) 2000 report, volume I: report to the general assembly, with scientific annexes-sources. United Nations

United Nations Scientific Committee on the Effects of Atomic Radiation (UNSCEAR) (1993) Sources and effects of ionizing radiation. United Nations, New York

Vargas-González HH, Méndez-Rodríguez LC, García-Hernández J, Mendoza-Salgado RA, Zenteno-Savín T, Arreola-Lizárraga JA (2016) Persistent organic pollutants (POPs) in populations of the clam Chione californiensis in coastal lagoons of the Gulf of California. J Environ Sci Health B 51(7):435–445

Wariaghli F, Tilghman-Sibille A, El Abidi A, El Hamri H, Fekhaoui M, Yahyaoui A (2013) Anguilla anguilla L.: evaluation of the degree of heavy metal contamination in the Sebou estuary and in Moulay Bousselham lagoon reserve (Morocco). Int J Aquat Sci 4(2):69–82

Yadav M, Kumar P, David A, Jindal MK (2025) Transportation of radioactivity in the marine world. In: Radiation status in the marine world. Springer, Cham, pp 95–131

Yang P, Hu Z, Shu Q (2021) Factors affecting soil organic carbon content between natural and reclaimed sites in Rudong Coast, Jiangsu Province, China. J Mar Sci Eng 9(12):1453

Zare MR, Mostajaboddavati M, Kamali M, Abdi MR, Mortazavi MS (2012) ^{235}U, ^{238}U, ^{232}Th, ^{40}K and ^{137}Cs activity concentrations in marine sediments along the northern coast of Oman Sea using high-resolution gamma-ray spectrometry. Mar Pollut Bull 64(9):1956–1961

Zhu YG, Shaw G (2000) Soil contamination with radionuclides and potential remediation. Chemosphere 41(1–2):121–128

Chapter 7
Microplastics: Transfer Pathways, Radiotracing and Challenges for Coastal Environments

Imane Bencheikh, Karima Azoulay, Khadija El-Moustaqim, Najat Qisse, Driss Azdem, and Jamal Mabrouki

Abstract Wastewater, industry, runoff, and degraded plastics pathways spread microplastics into waterways, and coastal areas. Lagoons, estuaries, and river systems, as semi-enclosed and ecologically sensitive environments, thus become prime reservoirs for their accumulation, increasing the risks to aquatic biodiversity and, indirectly, to human health. In this context, the International Atomic Energy Agency (IAEA) has, since 2024, strengthened its efforts through the NUTEC Plastics program, aimed at structuring an international network of laboratories, harmonizing analytical protocols, and developing next-generation detection technologies, ranging from satellite imagery integrated with artificial intelligence to spectroscopic approaches and advanced sensors. At the same time, nuclear techniques, and in particular radiotracing, are opening up new perspectives for monitoring in real time the transport, accumulation, and excretion of microplastics in aquatic and biological environments, through the use of radiolabeled plastics and non-destructive analyses. However, challenges remain, notably the stability of isotopes in saline and pH-varying environments, which can compromise the reliability of in situ measurements. This work highlights these recent advances and limitations, while emphasizing the potential of nuclear approaches, integrated with emerging technologies, to trace the fate of microplastics from urban effluents to sensitive coastal ecosystems. This synergy paves the way for more precise monitoring and a better understanding of the ecological and health impacts linked to plastic pollution.

I. Bencheikh (✉) · K. Azoulay · N. Qisse · D. Azdem · J. Mabrouki
Laboratory of Spectroscopy, Molecular Modeling, Materials, Nanomaterial, Water and Environment, CERNE2D, Faculty of Science, Mohammed V University in Rabat, Rabat, Morocco
e-mail: imane-bencheikh@um5r.ac.ma

K. El-Moustaqim
Improvement and Valuation of Plant Resources, Faculty of Sciences, Ibn Tofaïl University, Kenitra, Morocco

N. Mejjad (ed.), *Nuclear-Based Approaches application for marine pollution assessment and monitoring*, Environmental Science and Engineering,
https://doi.org/10.1007/978-3-032-12197-4_7

Keywords Microplastics · Radiotracing · Environmental pollution · Emerging contaminants

7.1 Introduction

The early twentieth century marked a major turning point in industrial history with the emergence of synthetic resins, which profoundly transformed materials science and manufacturing processes. The first polyester resins, used notably in coatings (Kim and Hamilton 2023), laid the foundations for a booming plastics industry. Today, more than 6000 grades of materials are available, suitable for uses as diverse as everyday packaging to the most sophisticated medical devices (Knapczyk et al. 1992). This expansion has led to global production approaching half a billion tons per year, dominated by polyethylene and polypropylene, which constitute the majority of industrial and consumer products (Skoglund et al. 2023). While this development has spurred innovation and facilitated modern life, it has also generated considerable environmental impacts, particularly through microplastic pollution. Microplastics, fragments resulting from the degradation of plastic or manufactured objects, are now recognized as one of the most concerning emerging contaminants for aquatic ecosystems. Transported primarily by wastewater, they largely escape conventional sanitation systems, which are unable to completely eliminate them (Skoglund et al. 2023). Their accumulation in natural environments disrupts biodiversity, threatens the health of aquatic organisms, and, through bioaccumulation, poses a risk to human health by contaminating marine and freshwater food chains essential to global resources. Coastal areas—rivers, estuaries, and lagoons—play a role as privileged receptacles, becoming reservoirs of accumulation that exacerbate these ecological and health risks.

However, monitoring microplastics remains complex. Their small size, diverse compositions, and variable behavior in heterogeneous environmental conditions limit the effectiveness of conventional analytical approaches. While recent advances have improved detection, these methods often lack the sensitivity and precision to provide real-time data. It is in this context that innovative approaches, particularly nuclear techniques such as radiotracing, are attracting growing interest. Using radiolabeled microplastics makes it possible to track their transport, accumulation, and interactions with organisms and ecosystems non-destructively and with high precision.

These advances are in line with recent efforts by the International Atomic EnergyAgency (IAEA), which launched the NUTEC Plastics program in 2024. The latter aims to structure an international network of laboratories, harmonize protocols, and develop new detection approaches, ranging from satellite imagery supported by artificial intelligence to next-generation sensors. At the same time, the application of radiotracing to coastal systems raises specific challenges, particularly related to isotopic stability in saline environments. Salinity and pH gradients can indeed alter the distribution and behavior of radioisotopes, compromising their reliability as tracers (Ramadan et al. 2014; Hou and Hou 2012). However, techniques such as accelerator mass spectrometry (AMS) can sensitively detect isotopes such as ^{14}C or ^{129}I, and the analysis of isotope ratios (e.g., $^{129}I/^{127}I$) provides crucial information on the origin and impact of pollution (Povinec 2000; Brauer and Ballou 1975). Despite these limitations, the use of such isotopes remains promising thanks to recent advances in analytical instrumentation and standardization efforts.

Thus, this review examines recent advances in nuclear techniques, and in particular radiotracing, for monitoring microplastics and emerging contaminants. It assesses their applicability, advantages and limitations, while highlighting their potential to fill current gaps in understanding the fate of microplastics and their environmental impacts.

7.2 Results and Discussion

7.2.1 Microplastics Pathways from the Management in the Purification Stations to Lagoons and Coastal Systems

Microplastic pollution is currently a major environmental problem, particularly visible in wastewater treatment plants (WWTPs), lagoons, and coastal areas. WWTPs represent a significant pathway for the introduction of microplastics into aquatic environments, as they are not always able to completely remove them from treated effluent or sludge. This difficulty is compounded by the diversity of microplastic sources—domestic activities, industrial discharges, and even urban runoff—and by the lack of standardized protocols for their detection and removal. Understanding the management of these particles in WWTPs and their downstream impacts on lagoons and coastal environments is therefore essential to assess and limit their dispersion. In WWTPs, microplastics originate from a multitude of sources, and their detection remains particularly complex due to their small size and heterogeneity, requiring the use of advanced analytical techniques (Gorfad et al. 2024; Mall and Solanki 2024). Current removal technologies rely primarily on membrane filtration or advanced oxidation processes, but these face significant constraints such as membrane fouling and high operating costs (Mall and Solanki 2024). While some WWTPs achieve reduction rates of up to 99%, efficiency varies greatly depending on the technology used. Tertiary treatments such as ultrafiltration show promising potential, but conventional systems still allow a significant fraction of particles to pass through (Fan et al. 2025; Sadia et al. 2022). Lagoons, which are a low-cost treatment solution, are much less effective at retaining microplastics. Operating conditions strongly influence their effectiveness, but their exact role in controlling these pollutants remains insufficiently studied (Fan et al. 2025). Downstream, discharges from WWTPs and lagoons directly contribute to coastal pollution, where microplastics accumulate and affect marine biodiversity and coastal ecosystems.

The inability of current systems to ensure complete elimination therefore increases pressure on coastal areas, highlighting the urgency of appropriate management strategies.

The results of the study conducted at the wastewater treatment plant in Boulogne-sur-Mer clearly illustrate this paradox. On the one hand, the STP acts as an effective barrier, retaining nearly 98% of microplastics thanks to the successive stages of pretreatment, primary settling, and biological treatment. On the other hand, due to the considerable volumes of water treated daily, it remains a vector of transfer: several million residual particles still reach waterways and, by extension, the surrounding lagoons and coastal areas (Kouidri 2022).

Beyond their accumulation in sediments and the water column, these microplastics can become a chronic source of pollution for surrounding lagoons, ecosystems already weakened by eutrophication and human activity. Their presence poses a threat to benthic fauna, filter-feeding organisms, as well as to human communities dependent on local fisheries (Kouidri 2022; Amara and Kazour 2020).

Thus, wastewater treatment plants, while playing an essential filtering role, are not sufficient to stop the spread of microplastics. To limit their transfer to lagoons and coastal areas, several complementary levers are needed:

- Improving tertiary treatment processes (membrane filters, advanced oxidation processes).
- Reduction at source (banning intentionally added microplastics, designing lower-emission textiles).
- Implementing specific regulations integrating microplastics into water quality standards enhancing the use of radiotracing raw materials.

7.2.2 *Nuclear Techniques in Environmental Monitoring: Microplastics Case*

The use of nuclear techniques for environmental monitoring has a history spanning several decades, marked by atmospheric nuclear tests, the growth of nuclear power plants, and major accidents such as Chernobyl. These techniques offer measurement and analysis tools capable of detecting radionuclides at very low concentrations, thus making it possible to monitor their dispersion in air, water, soil, and living organisms. In France, the French Institute for Radiological Protection and Nuclear Safety (IRSN) has gradually developed a range of systems combining spot sampling and automated remote monitoring networks. Stations spread across the country take more than 30,000 samples annually (aerosols, soil, food, water, bioindicators), supplemented by laboratory analyses using gamma spectrometry, tritium and carbon-14 measurement, and thermoluminescent dosimetry (Frère 2017).

Automated networks particularly illustrate the contribution of nuclear technologies. For example, TELERAY enables continuous measurement of the ambient gamma dose rate via 164 beacons deployed in mainland France and overseas, while

SARA characterizes aerosol radioactivity in near real-time. Similarly, HYDROTELERAY monitors the main rivers downstream of nuclear facilities using gamma spectrometry, and TELEHYDRO focuses on gamma-emitting radionuclides in urban wastewater treatment plants (Frère 2017).

These systems serve two complementary functions:

- A monitoring function, aimed at rapidly detecting any abnormal increase in radioactivity and triggering an alert.
- A characterization function, allowing the assessment of chronic exposure of populations and ecosystems in normal or accidental situations.

The effectiveness of these techniques is measured by their ability to provide long-term and comparable chronicles over time. The series of measurements of cesium-137 in the air between 1958 and 2004 constitute a valuable example, making it possible to trace the impact of nuclear tests and the gradual decline in activity levels (Russell et al. 2015).

Current prospects call for modernizing these networks in the face of aging equipment, while integrating more sensitive technologies capable of measuring difficult-to-detect radionuclides such as tritium or certain long-lived beta emitters. The IRSN is therefore undertaking renovation efforts (SARA-II, mobile beacons) to improve real-time detection and strengthen resilience in the event of a major accident.

In the case of microplastics, radiotracing offers the possibility of monitoring the fate of pollutant particles in an aquatic or treatment system in real time, providing unique information on residence times, transfer rates, and retention mechanisms. Nuclear spectrometry techniques (gamma, alpha) make it possible to detect the presence of radioactive elements associated with industrial or medical waste at very low concentrations.

Figure 7.1 illustrates three main strategies used for radiolabeling microplastics, each addressing specific environmental monitoring and analysis objectives.

The first approach is based on the direct incorporation of radionuclides (e.g., ^{51}Cr, ^{125}I, ^{85}Sr, or ^{46}Sc) into a plastic matrix, typically polystyrene. This simple and effective method has shown good results for short-term studies. However, it has a major limitation: the risk of gradual release of radioactive isotopes, which reduces its reliability for long-term experiments (Lanctôt et al. 2018).

To overcome this problem, a second approach has been developed: the encapsulation of a radioactive nanoscale core, such as labeled silver (^{110m}Ag), in a polymer shell. This process creates a protective barrier that prevents the direct dispersion of the radionuclide into the environment, while maintaining the ability to accurately track particles.

Finally, the third method involves the neutron activation of a metal core encapsulated in a polymer. Exposed to a neutron flux in a reactor, this core becomes radioactive and emits a measurable signal. This technique produces stable and traceable particles without significant release, thus providing a robust solution for long-term studies.

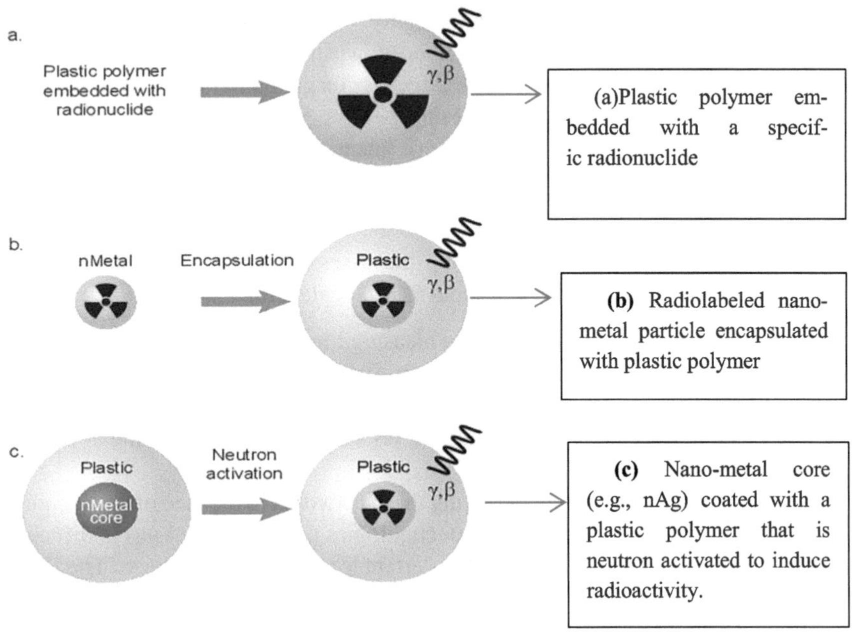

Fig. 7.1 Microplastics radiotracing types: Schematic representation (Lanctôt et al. 2018)

By combining these approaches, the figure highlights the evolution of radiolabeling strategies, which aim to improve the reliability, sensitivity, and safety of environmental analyses on the fate of microplastics.

7.2.3 Innovations, Challenges and Perspectives of Nuclear Techniques Applied for Coastal System Monitoring

Nuclear techniques offer innovative tools for marin ecosystem monitoring of emerging pollutants such as microplastics. Unlike traditional, often limited approaches, radiotracing using isotopes such as iodine-131 or zinc-65 allows for real-time monitoring of the movement of these particles in complex environments, without altering sample integrity (Fara and Muslim 2024; Munir et al. 2024). These techniques offer the ability to measure key parameters such as residence times, transfer rates, and retention rates, thus providing essential data for hydrodynamic modeling and the validation of remediation processes. Furthermore, nuclear spectrometry methods (gamma, alpha) enhance the sensitivity of analyses, enabling the detection of minute concentrations often inaccessible by conventional tools.

However, these advances are not without challenges. Their implementation requires specialized infrastructure, highly qualified personnel, and strict compliance with radiation protection standards. Investment costs and regulatory complexity still limit their dissemination, particularly in contexts where analytical capacities are limited. Furthermore, the risk of alteration of particle properties during isotopic labeling remains a methodological challenge that must be managed to ensure the ecological representativeness of the results (Suseno et al. 2023).

In addition, while radiotracing techniques offer unprecedented potential for monitoring the dispersion and transformation of microplastics, their application in marine and coastal environments remains hampered by several scientific and technical challenges (Table 7.1). The first concerns the stability of radioactive isotopes under complex physicochemical conditions. Saline environments, characterized by high ion concentrations (Na^+, Cl^-, Mg^{2+}, SO_4^{2-}), and pH variations directly affect the sorption and desorption mechanisms of radionuclides on plastic surfaces. These processes can lead to rapid isotope redistribution, thus compromising the reliability of in situ measurements and the accurate reconstruction of transport trajectories (Cresswell et al. 2020; Yadav et al. 2024). A second major challenge is related to the formation of biofilms on microplastics exposed to the marine environment. The bacteria, algae, and organic particles that colonize these surfaces profoundly alter their charge, roughness, and density. While this colonization tends to increase radionuclide retention, it also causes a buoyancy transition (from positive to negative), promoting particle sedimentation and blurring the recorded isotopic signals (Yadav et al. 2024). Isotopes can then be trapped heterogeneously, with variable transfer kinetics depending on the plastic's degradation stages. Added to this are analytical

Table 7.1 Main challenges of radiotracing microplastics in marine and coastal environments (Cresswell et al. 2020; Yadav et al. 2024)

Main challenge	Description	Impact on the measures
Isotopic stability in salt water	Na^+, Cl^-, Mg^{2+} ions and carbonates complex radioactive isotopes (Cs, Sr, etc.)	Rapid desorption, loss of radiotracer signal, reduced in situ reliability
pH variability	pH fluctuations linked to river inputs, tides, eutrophic zones	Alters the chemical speciation of radionuclides, distorts the distribution coefficients
Formation de biofilms	Colonization by bacteria, algae, protozoa on microplastics	Increases the adsorption of isotopes but modifies the buoyancy (from floating to sedimented)
Degradation of plastics	Aging, cracking and surface oxidation (UV, radiation, mechanical abrasion)	Makes adsorption heterogeneous and variable over time
Faibles coefficients de distribution (Kd)	Adsorption of Cs and Sr on MP significantly lower than that on sediments (2–3 orders of magnitude)	Makes quantitative interpretation and modeling of flows difficult
Variabilité hydrodynamique	Coastal currents, tides, upwellings and river inputs	Create local micro-conditions that amplify the uncertainty of isotopic monitoring

and methodological constraints. In real-life conditions, isotopic signals are diluted in complex matrices, subject to fluctuations in temperature, oxygenation, and redox. Unstable isotopes, sensitive to complexation with carbonates, sulfates, or dissolved organic matter, risk losing their original "fingerprint." Furthermore, the distribution coefficient values observed for Cs-137 and Sr-85 on microplastics are two to three orders of magnitude lower than those measured on marine sediments, which complicates quantitative interpretation and can introduce biases when modeling fluxes (Yadav et al. 2024). Finally, the spatial and temporal heterogeneity of coastal environments amplifies these uncertainties. Tides, longshore currents, river inputs, and upwelling phenomena create a mosaic of microconditions that can accelerate desorption or, conversely, stabilize certain isotopes. This variability makes it difficult to extrapolate results obtained in the laboratory or mesocosm to the scale of natural ecosystems (Cresswell et al. 2020).

Future prospects lie in the development of safer isotopic tracers (short half-life or encapsulated), the integration of nuclear approaches with complementary tools (Raman spectroscopy, hyperspectral imaging, numerical modeling), and the use of artificial intelligence to interpret the large amounts of data generated. Such synergy would enable a shift from one-off studies to continuous, multi-scale monitoring, improving our understanding of pollutant dynamics and facilitating the development of more effective management strategies.

Ultimately, nuclear techniques, although costly and technically demanding, represent a strategic asset for filling gaps in environmental research.

7.3 Conclusion

Nuclear techniques offer innovative tools for environmental monitoring of emerging pollutants such as microplastics. Unlike traditional, often limited approaches, radiotracing using isotopes such as iodine-131 or zinc-65 allows for real-time monitoring of the movement of these particles in complex environments, without altering sample integrity. These approaches pave the way for a better understanding of transport, bioaccumulation, and degradation mechanisms. At the same time, nuclear magnetic resonance (NMR) provides a non-destructive view of the biochemical processes influenced by these contaminants, while atomic spectrometry, thanks to its high sensitivity, improves the detection of microplastics in various matrices. However, these innovations come with challenges, whether it is the need to strengthen metrological capabilities, reduce analytical costs, or integrate these tools into still-incomplete regulatory frameworks. The outlook is nevertheless promising: by combining radiotracing, spectrometry, and cutting-edge technologies, nuclear techniques could permanently transform environmental monitoring and contribute to more effective management of pollution risks.

References

Amara R, Kazour M (2020) Les microplastiques au sein des stations d'épuration : flux, élimination et apports aux milieux aquatiques. Cas de la station d'épuration Séliane de Boulogne-sur-Mer. Laboratoire d'Océanologie et Géosciences (LOG-UMR 8187) de l'Université du Littoral Côte d'Opale

Brauer FP, Ballou NE (1975) Rapports isotopiques de l'iode et d'autres radionucléides comme indicateurs de pollution nucléaire. http://www.osti.gov/scitech/biblio/4160280-isotopic-ratios-iodine-other-radionuclides-nuclear-power-pollution-indicators

Cresswell T, Metian M, Fisher NS, Charmasson S, Hansman RL, Bam W, Bock C, Swarzenski PW (2020) Exploring new Frontiers in marine radioisotope tracing—adapting to new opportunities and challenges. Front Mar Sci 7:406. https://doi.org/10.3389/fmars.2020.00406

Fan L, Gao L, Mohseni A (2025) Devenir et contrôle des microplastiques dans les systèmes de traitement des eaux usées municipales. In: Microplastic pollution, pp 54–67. https://doi.org/10.1201/9781032706573-4

Fara M, Muslim M (2024) Application de l'analyse nucléaire à la bioaccumulation de microplastiques avec de l'iode 131 dans les organismes marins. Revue Mondiale des Sciences et Technologies de l'Ingénierie Avancées 12(1):446–453. https://doi.org/10.30574/wjaets.2024.12.1.0251

Frère L (2017) Les microplastiques: une menace en rade de Brest? l'Université de Bretagne Occidentale (UBO), Laboratoire des sciences de l'Environnement MARin (LEMAR)

Gorfad RS, Ribadiya HN, Rathod VK, Parmar MR, Mayani SV (2024) Étude des approches et technologies pour les microplastiques dans la gestion des eaux usées. pp 90–106. https://doi.org/10.1201/9781003438793-7

Hou X, Hou Y (2012) Analyse de l' iode 129 et son application comme traceur environnemental. J Anal Sci Technol 3(2):135–153. https://doi.org/10.5355/JAST.2012.135

Kim S-T, Hamilton R (2023) Polymères synthétiques: polyesters, polyéthers, polysulfones et autres polymères. In: Toxicologie de Patty. Wiley, pp 1–40. https://doi.org/10.1002/0471125474.tox092.pub3

Knapczyk JK, Robert S, Melvin H (1992) Basic concepts of extrusion. In: Résines synthétiques et plastiques, pp 623–707. https://doi.org/10.1007/978-1-4757-6431-4_19

Kouidri R (2022) Niveau de contamination des matrices environnementaux par les microplastiques: review de la litérrature. Faculté des Mathématiques et des Sciences de la Matière, Université Kasdi Merbah—Ouargla

Lanctôt CM, Al-Sid-Cheikh M, Catarino AI, Cresswell T, Danis B, Karapanagioti HK, Mincer T, Oberhänsli F, Swarzenski P, Tolosa I, Metian M (2018) Application of nuclear techniques to environmental plastics research. J Environ Radioact 192:368–375. https://doi.org/10.1016/j.jenvrad.2018.07.019

Mall C, Solanki P (2024) Défis et devenir des microplastiques dans les procédés de traitement des eaux usées. In: Water challenges of an urbanizing world. IntechOpen, pp 201–211. https://doi.org/10.1201/9781003486947-10

Munir M, Subcchi M, Nurmanjaya A, Prasetya KE, Rindiyantono F, Chairuman, Pratama C, Yanto, Pujiyanto A, Setiawan H, Sarwono DA, Sarmini E, Fara ME, Suseno H (2024) Développement d'un modèle microplastique à base de polystyrène pour l'étude de la bioaccumulation et de la biodistribution à l'aide de la méthode de radiotraçage et d'analyse nucléaire. Mar Pollut Bull 201:116283–116283. https://doi.org/10.1016/j.marpolbul.2024.116283

Povinec PP (2000) Mesures AMS de ^{14}C et ^{129}I dans l'eau de mer autour des sites de décharge de déchets radioactifs. Nuclear Instrum Methods Phys Res 172:672–678

Ramadan KA, Seddeek MK, Sharshar T, Elnimr T, Badran HM (2014) Effet de la salinité sur les concentrations de radio-isotopes dans le milieu aquatique d'une lagune côtière hypersalée. Isot Environ Health Stud 50(2):235–242. https://doi.org/10.1080/10256016.2014.870566

Russell BC, Croudace IW, Warwick PE (2015) Determination of ^{135}Cs and ^{137}Cs in environmental samples: a review. Anal Chim Acta. https://doi.org/10.1016/j.aca.2015.06.037

Sadia M, Mahmood A, Ibrahim M, Irshad M, Quddusi AHA, Bokhari A, Mubashir M, Chuah LF, Show PL (2022) Pollution par les microplastiques des stations d'épuration: analyse critique des défis, de la détection, des techniques d'élimination durable et de l'économie circulaire. Environ Technol Innov 28:102946. https://doi.org/10.1016/j.eti.2022.102946

Skoglund R, Le Roy R, Cavender FL (2023) Polymères synthétiques. In: Toxicologie de Patty. Wiley, pp 1–15. https://doi.org/10.1002/0471125474.tox088.pub3

Suseno H, Muslim M, Munir M, Pujiyanto A, Yusof NF (2023) Microplastic bioaccumulation by tiger snail (Babylonia spirata): application of nuclear technique capability using polystyrene labelled with radiotracer ^{65}Zn. Ilmu Kelautan Indones J Mar Sci 28(1):81–89. https://doi.org/10.14710/ik.ijms.28.1.81-89

Yadav S, Rout S, Joshi V, Pulhani V, Kumar AV (2024) Unfolding the interaction of radioactive Cs and Sr with polyethylene-derived microplastics in marine environment. Discov Oceans 1:13. https://doi.org/10.1007/s44289-024-00015-8

Chapter 8
Nuclear and Isotopic Methods for Quantifying Blue Carbon Stocks, Estimating Sequestration Rates, and Identifying Sources

Nezha Mejjad, Abdelmourhit Laissaoui, and Moncef Benmansour

Abstract Nuclear and isotopic approaches are widely used in the field of environmental science as powerful tools for mitigating and managing environmental issues. These approaches are employed to track the history of pollutants over time and identify their potential sources, which helps in establishing mitigation and remediation strategies. In blue carbon sciences, these approaches are indispensable. By applying radiometric dating, it becomes possible to track changes in carbon accumulation over time and to quantify carbon burial rates. This information is crucial for understanding the sequestration and storage capacities of blue carbon ecosystems, which, in turn, helps in restoration projects and management programs of coastal blue carbon ecosystems. In this sense, the present chapter reviews and analyzes the role of nuclear approaches in advancing knowledge in the field of blue carbon sciences.

Keywords Blue carbon · Nuclear approaches · Radioisotope · Stable isotope · Burial rate · Radiometric dating · Carbon storage

8.1 Introduction

Coastal Blue Carbon Ecosystems (CBCEs) play a crucial role in climate change mitigation, serving as effective nature-based solutions. These ecosystems, including seagrasses, mangroves, and salt marshes, provide numerous goods and services, such as the storage and sequestration of CO_2 (Siikamäki et al. 2012a; Daba and Dejene 2018; Mejjad et al. 2025). There is an increasing trend of interest in understanding their ecological role and their long-term capacities and resilience for CO_2 storage.

N. Mejjad (✉) · A. Laissaoui · M. Benmansour
National Centre for Nuclear Energy, Science and Technology (CNESTEN), Rabat, Morocco

N. Mejjad (ed.), *Nuclear-Based Approaches application for marine pollution assessment and monitoring*, Environmental Science and Engineering,
https://doi.org/10.1007/978-3-032-12197-4_8

The intensification of climate change effects necessitates exploring the capacities of these ecosystems for carbon storage as a means of sustaining this valuable regulatory service, thereby conserving and protecting them from anthropogenic and environmental pressures. To support this, analytical and standardized protocols were developed to measure carbon content, quantify stocks, and estimate the long-term sequestration rates (Goulden et al. 1996; Qureshi et al. 2012; Petrokofsky et al. 2012; Smith et al. 2020).

Nuclear and isotopic approaches are considered valuable tools for research and studies concerning blue carbon quantification and assessment. The use of nuclear methods has enabled the quantification of carbon accumulation rates over time (Arias-Ortiz et al. 2018). These approaches have also allowed the establishment of the history of carbon accumulation and the analysis of carbon concentration over time (Martin 2022).

Sediments accumulating in Coastal Blue Carbon (CBC) ecosystems—such as mangrove forests, seagrasses, and salt marshes—can be analyzed to reconstruct environmental changes spanning from recent decades to millions of years. At the same time, the capacity of these vegetated coastal ecosystems to sequester and store carbon within their sediments can be quantified using isotopic and nuclear techniques (Arias-Ortiz et al. 2018; Martin 2022).

Lead-210 and Caesium-137 were widely used in marine pollution monitoring projects. These radiotracers trace the history of sediment accumulation rates and pollutant deposition, which helps define the possible sources of pollution. Similarly, for blue carbon, ^{210}Pb and ^{137}Cs are considered critical tools for blue carbon ecosystem research as they allow sediment layer dating and estimation of carbon accumulation rates over time, where ^{210}Pb acts as a natural chronometer, allowing the dating of sediment cores over the past 100–150 years (Arias-Ortiz et al. 2018; Schirone et al. 2022). By measuring the activity concentration of lead-210, it is possible to calculate the sediment accumulation rates and the amount of trapped organic carbon. Decreased ^{210}Pb activities with depth deliver a carbon burial timeline of up to 100 years (IAEA 2021). Caesium is an anthropogenic and temporal marker released during atmospheric nuclear testing, with peaks in 1963 and Chernobyl in 1986. The peak of the artificial radionuclide is used to validate the dates established by ^{210}Pb and ensure accuracy in sediment dating.

Identifying the origin of carbon in coastal vegetated ecosystem sediment is crucial in blue carbon science, as it provides knowledge related to changes in carbon sources, which allows for an understanding of ecosystem dynamics and the distinction between autochthonous and allochthonous carbon. This distinction is crucial for accurately calculating net ecosystem carbon sequestration and thus assessing the real contribution of blue carbon ecosystems to climate change mitigation. Stable isotopes such as δ^{13}C and δ^{15}N are widely used for this purpose. Tang et al. (2023), by measuring TOC, TN, δ^{13}C, and δ^{15}N in two sediment cores collected from a mangrove ecosystem in China, reported that algae and mangrove plants are the main sources of organic matter, where mangrove-derived organic carbon contributed >50% of OM in numerous mangrove-dominated sediment sites. Such data are critical for blue carbon credit schemes, as only carbon confirmed to be derived from blue carbon ecosystems can be counted toward their climate value. This chapter examines the contributions of nuclear and isotopic approaches to the field of blue carbon research and reviews the current understanding of blue carbon ecosystems.

8.2 Blue Carbon Sequestration: State of Knowledge

The blue carbon concept was first introduced in 2009 and refers to carbon captured and stored in coastal and marine environments (Nellemann et al. 2009; Lovelock and Duarte 2019). The role that marine and coastal ecosystems play as a sink for CO_2 was previously known, but the concept of "blue carbon" has further helped highlight it and make it more visible to various communities, including the scientific community and the coastal community. Thus, Blue Carbon Coastal Ecosystems, mainly mangroves, seagrasses, and salt marshes, are now recognized for their vital role in mitigating climate change.

As shown in Fig. 8.1, the term "Blue carbon" was first introduced in 2009 in a UNEP/FAO report. However, several studies conducted between 1970 and 1990 have indirectly highlighted the importance of marine and coastal ecosystems and their capacity for carbon storage (Zieman 1982; Stevenson 1988; Zieman and

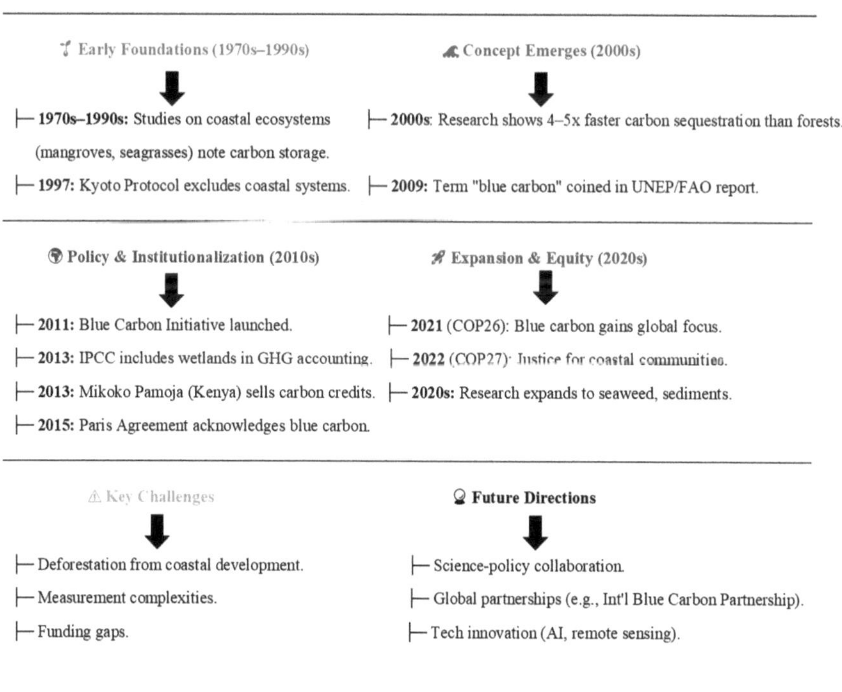

Fig. 8.1 Chronology of Blue Carbon foundation and development

Zieman 1989). In 1997, the Kyoto Protocol was adopted, an international agreement aimed at reducing greenhouse gas (GHG) emissions, primarily in industrialized countries and economies in transition (Encyclopædia Britannica). However, CBC was excluded from the agreement. The concept of coastal systems' role in GHG emissions mitigation started emerging in the 2000s, while the term "Blue Carbon" was first introduced in 2009 (Nellemann et al. 2009; Schrag 2009). In 2010, the institutionalisation of the blue carbon as a solution for climate change was known, while in 2011, a collaborative effort by IUCN, Conservation International, and IOC-UNESCO, known as "the Blue Carbon Initiative" (https://www.thebluecarboninitiative.org/), was launched to promote research, advance policy frameworks, and projects related to coastal ecosystems. 2013 was marked by the release of the IPCC (Intergovernmental Panel on Climate Change) Wetlands Supplement. The latter provided standardized guidelines to incorporate wetlands, mangroves, salt marshes, and seagrasses into nations' GHG inventories for stronger carbon accounting.

Blue carbon has received formal recognition in the global climate policy by 2015 through the Paris Agreement, acknowledging the important role of oceans and coastal environments in climate change mitigation-related strategies. This led countries to include the restoration and conservation of blue carbon ecosystems in their nationally Determined Contributions (NDCs). Indeed, 45 countries integrated coastal and marine NbS for mitigation and adaptation purposes, 25 for adaptation only, and 1 for mitigation only (Blue Carbon Initiative 2021). By 2021, blue carbon had received more attention in global climate negotiations, particularly at COP26 (2021), underlining its crucial role in both mitigation and adaptation. At COP27 (2022), the ocean-climate linkage was prioritized, highlighting the need for justice in supporting communities reliant on coastal ecosystems (Morgera and Lennan 2024).

Despite the growing interest and recognition of the role of blue carbon ecosystems in mitigating and adapting to climate change, these ecosystems continue to face numerous challenges. Deforestation caused by coastal development, aquaculture farming, and climate change degrades seagrasses and mangroves, potentially leading to the release of stored carbon (Ahmed and Glaser 2016). Additionally, blue carbon faces methodological challenges, leading to inaccuracies in the estimation of blue carbon (Serrano et al. 2023). These challenges necessitate collaboration between science and policy to refine methods and techniques related to blue carbon estimates.

8.3 Blue Carbon Ecosystems

8.3.1 Seagrass

Seagrass beds are known as 'ecosystem engineers' because they play a crucial ecological role and offer numerous beneficial ecosystem services to marine biodiversity and human populations. By definition, seagrass beds consist of aquatic flowering plants (angiosperms) that have colonized the oceans since the time of the dinosaurs (James and Larkum 1996; Den Hartog and Kuo 2006) and are the only type of flowering plants capable of living entirely underwater. Seagrass beds thrive in sediments at the ocean's bottom, featuring upright, elongated leaves and a buried root structure (rhizome), as illustrated in Fig. 8.2.

Seagrass beds are found in the coastal waters of most continents. They are the main food source for dugongs and green turtles and serve as habitats for many small marine animals, some of which are commercially important, such as shrimp and fish. They also contribute to water clarity by absorbing nutrients from coastal runoff and stabilizing sediments.

Seagrass beds are composed of flowering plants (not algae). Approximately 70 species of seagrass have been described worldwide, distributed across 12 genera, 4 families (*Posidoniaceae*, *Zosteraceae*, *Cymodoceaceae*, and *Hydrocharitaceae*), and 2 orders. They can form true underwater meadows. They grow at shallow depths where light is sufficient for photosynthesis to occur. Their entire life cycle takes place underwater, including pollination.

These marine flowers play an essential role in carbon capture and storage. Seagrasses sequester CO_2 through photosynthesis and store it in their biomass (roots, leaves, rhizomes) and sediments for centuries (Howard et al. 2014). They capture carbon 35 times faster than tropical rainforests (Fourqurean et al. 2012; Singh et al. 2022), accounting for 10–18% of the total carbon storage by the ocean each year (Singh et al. 2022), despite covering less than 0.1% of the seafloor.

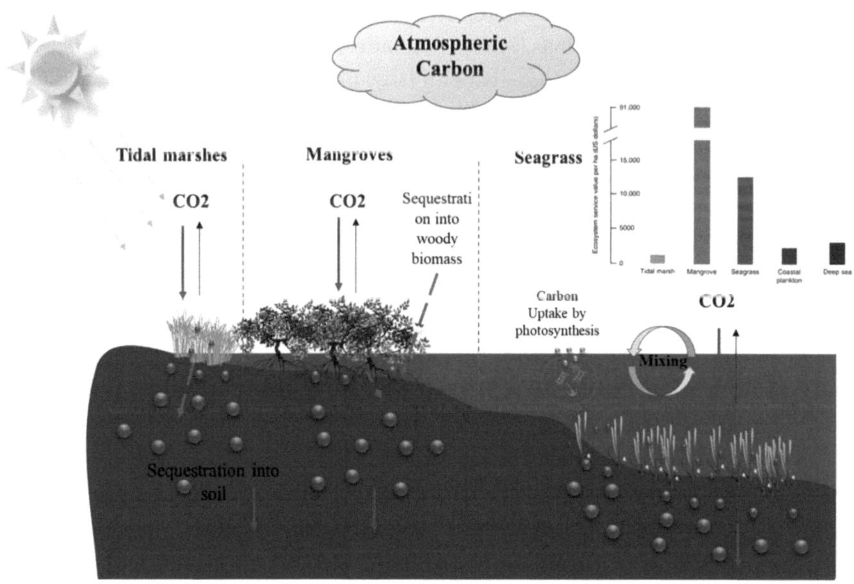

Fig. 8.2 Three blue carbon ecosystems

8.3.2 Mangroves

Mangroves are salt-tolerant trees and shrubs that grow along the marine shoreline in subtropical and tropical regions. They cover around 152,000 km^2, which is equivalent to 0.04% of the surface of Earth (UNEP 2014). Mangrove forests are found in 123 countries where over 100 million people live within 10 km of these ecosystems. They provide numerous ecosystem services and are beneficial to people living around them, including forest products, fisheries, and clean water, and play an essential role in protecting the coastline against erosion, storms, and extreme weather events (UNEP 2014).

Besides, Mangroves help mitigate the effects of climate change by capturing and storing carbon in their biomass and soils. CO_2 is captured through photosynthesis. They are known as blue carbon ecosystems. According to (Siikamäki et al. 2012b), around 11.5 billion tons of carbon are stored by blue ecosystems (Seagrass, Mangrove, and salt marsh), with 6.5 billion tons stored by mangroves, representing the largest pool of blue carbon. Consequently, mangroves are important for mitigating the effects of climate change, with significant sequestration rates and long-term storage of carbon.

8.3.3 Salt Marsh

Salt marshes are dominated by halophytic, mainly herbaceous plants and shrubs. They are found in areas such as lagoons, estuaries, deltas, and tidal creeks. It covers around 5.5 million hectares globally, with large extents along the Atlantic and Gulf coasts of North America, followed by Northern Europe, Southern Australia, and East Asia (Murray et al. 2011; Campbell et al. 2022; Worthington et al. 2024). Salt marshes provide many ecosystem services, including carbon sequestration and storage in vegetation and soils. If left undisturbed, these intertidal wetlands can retain carbon for millennia, making them crucial for mitigating the effects of climate change.

8.4 The Use of Nuclear Approaches for Blue Carbon Sequestration Quantification

The quantification of carbon sequestration rates in blue carbon ecosystems necessitates robust and sophisticated methods to evaluate the magnitude and the stability of stored carbon in sediments over time. Isotopic and nuclear approaches have emerged as powerful and complementary tools to traditional geochemical and ecological methods. Isotopic and nuclear approaches enable the determination of sedimentation rates, carbon burial efficiencies, the history of carbon content and stocks over time, and the long-term stability of organic matter under changing environmental conditions.

8.4.1 *Sediment Dating and Accumulation Rates*

Organic carbon accumulated in sediment is the biologically derived fraction, including algae, plants, microbes, and detritus, which are preserved and buried in sedimentary environments. This carbon can have autochthonous and/or allochthonous sources. OM settles through the water column as particulate organic carbon (POC) or is dissolved as dissolved organic carbon (DOC) that later adsorbs onto mineral particles (Fig. 8.3). For burial to occur, deposition must exceed decomposition. Various factors can influence the preservation and burial of organic carbon, including redox conditions, mineral association, and sedimentation rate.

Radionuclides such as Pb-210 and Cs-137 are widely used for reconstructing and establishing chronologies of sediment over the past 100–150 years (Mejjad et al. 2016, 2018, 2022, 2023). Additionally, the use of Pb-210 enables the calculation of sedimentation rates and sediment accumulation rates over several years (Mejjad et al. 2022). These profiles of sedimentation rates are crucial for calculating carbon

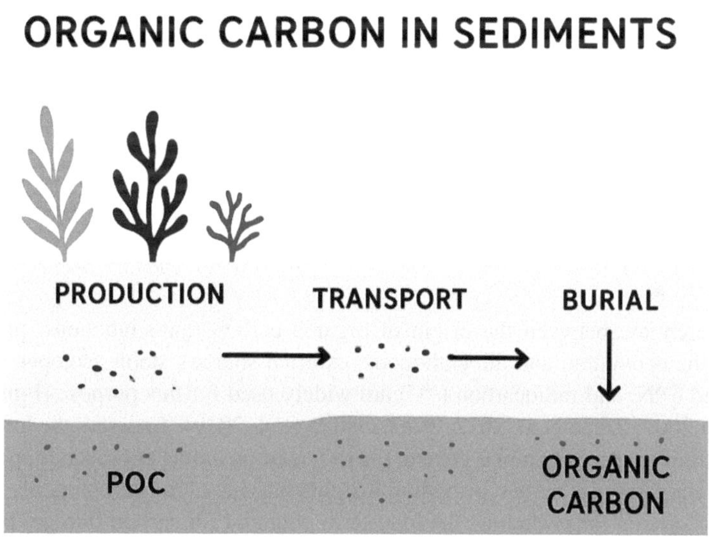

Fig. 8.3 Organic carbon stored in sediment

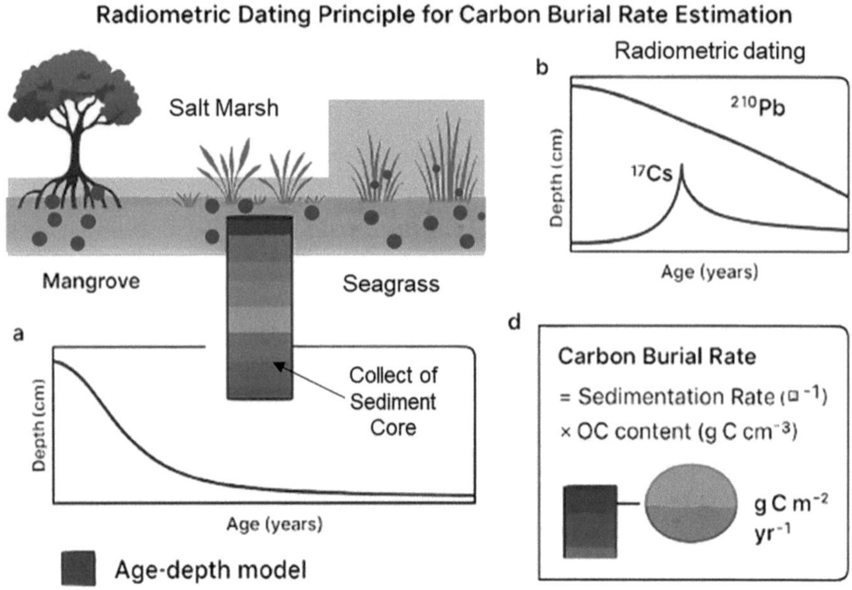

Fig. 8.4 The use of radiometric dating for age establishment and carbon accumulation rate determination

accumulation rates. Besides, the ages provided by Pb-210 and Cs-137 dating allow for establishing the history of carbon deposition, stock, and carbon accumulation rates over the past 100–150 years. Thus, by combining OC content measurement with calculated sedimentation rates in sediment core layers, the annual carbon burial rates can be estimated with precision (Fig. 8.4). This would help track changes in carbon content and carbon accumulation rates over the years.

8.4.2 *Tracing Carbon Sources and Stability Using Stable Isotopes ^{13}C and ^{15}N*

To differentiate between the origin of organic carbon (autochthonous: produced within the ecosystem and allochthonous: external source), stable isotopes such as $\delta^{13}C$ and $\delta^{15}N$, and radiocarbon (^{14}C) are widely used for this purpose (Finlay and Kendall 2007; Zigah et al. 2012, 2017; Garcia et al. 2019). To assess the long-term potential of sequestration in a system, the radiocarbon dating approach is applied to sedimentary organic matter, providing insights into how long carbon pools persist, which is crucial for evaluating the long-term potential for carbon storage. Besides, information on carbon sources and fluxes among land, coastal, and marine ecosystems is crucial for stakeholders and managers to achieve ecosystem management goals and to support the role of blue carbon ecosystems in carbon capture and storage.

8.4.3 Stable Carbon Isotope Ratios ($\delta^{13}C$)

Stable carbon isotope ratios refer to the relative abundance of the heavier isotope (^{13}C) over the lighter one (^{12}C) in a sample. This ratio is standardized against an international reference (Vienna Pee Dee Belemnite, VPDB). This later expresses the zero point of the stable carbon isotope scale that is used to define the relative abundance of ^{13}C and ^{12}C (Coplen 2011).

As chemical, biological, and physical processes fractionate ^{13}C and ^{12}C differently, $\delta^{13}C$ allows providing information related to carbon source pathways and transformation. For instance, phytoplankton, mangrove salt marsh, and seagrass vegetation are primary producers with varied $\delta^{13}C$ values, which depend on factors related to their photosynthetic pathways and inorganic carbon utilization. Accordingly, this marine vegetation's leaves have distinct isotopic fingerprints, which make it possible to recognize the origin of organic carbon deposited in sediments (Geraldi et al. 2019). Besides, the value of $\delta^{13}C$ preserved in sediment cores can provide insight into changes in carbon sources over time, allowing for the establishment of the history of carbon burial and ecosystem dynamics.

8.4.4 Stable Nitrogen Isotope Ratios ($\delta^{15}N$)

$\delta^{15}N$ in the blue carbon ecosystem provides complementary information to $\delta^{13}C$, allowing for the tracing of nutrient inputs. High values of $\delta^{15}N$ in blue carbon ecosystems (seagrasses, mangrove, and salt marsh) reflect anthropogenic nutrient loading, which influences ecosystem health (Samper-Villarreal et al. 2020; Ke et al. 2020) and carbon storage capacity (Liu et al. 2023; Duarte et al. 2018; Jones et al. 2018). Additionally, $\delta^{15}N$ allows distinction between nitrogen originating from anthropogenic inputs, such as wastewater and agricultural runoff, and from natural marine processes, including upwelling and N_2 fixation. This distinction is important because the sources of nitrogen affect primary production and, consequently, the fate of stored carbon.

Isotopic data incorporate with nuclear and geochemical approaches to provide insight into coupled carbon-nitrogen biogeochemical cycles within the blue carbon. These combined methods help to evaluate decomposition rates, microbial processes, and the carbon stocks resilience under anthropogenic stressors and changing environmental conditions. Besides, the temporal variations of isotopic signatures in sediments can reveal whether carbon is being regularly retained or is undergoing transformations and export (The Information Architects of Encyclopaedia Britannica 2025). Stable isotopes thus serve as indirect indicators of long-term carbon sequestration efficiency.

8.5 Discussions

Nuclear and isotopic approaches are of great interest as they help in understanding carbon burial rates and factors that may affect carbon sequestration and storage efficiency. Such data are critical for restoration programs and management projects, particularly in the face of increasing natural and anthropogenic stressors that affect blue carbon ecosystems. Thus, identifying sources, quantifying the carbon accumulation rate, and reconstructing the history of carbon deposition would inform policymakers and managers to accurately restore and manage blue carbon ecosystems.

Nuclear approaches offer several strengths in the field of blue carbon, such as source tracing, quantification of burial rates, and impact reconstruction, which no other techniques can match. These include high-resolution sediment dating, which allows for reconstructing carbon burial history and rates, thereby enabling the reconstruction of climate and pollution impacts. Long-term archives from isotopes permit uncovering of signals of pollution, the history of eutrophication, and the resilience of blue carbon ecosystems under various natural and anthropogenic stressors.

However, these approaches have some limitations, including the need for radiation safety protocols and specialized laboratories. Cost and accessibility are among the main limitations and challenges of the use of nuclear techniques in blue carbon sciences. Table 8.1 below outlines the primary challenges associated with the application and use of nuclear approaches in blue carbon sciences.

As shown in Table 8.1, various challenges are associated with the radioanalytical techniques used in blue carbon sciences, including cost, infrastructure, and gaps in geographic coverage. To assess the geographic coverage of studies on blue carbon that employed nuclear approaches, primarily radiometric dating using Pb-210, we conducted a search in the Scopus database using the query "Blue carbon AND ^{210}Pb". Between 2001 and 2025, 63 documents were published, with almost all documents being published by institutes from developed countries. Australia ranked first with 24 documents, followed by the United States, Spain, and China, reflecting an unbalanced distribution and coverage of studies that utilized nuclear approaches in blue carbon sciences. This indeed supports the described challenges related to the use of radioanalytical methods.

Figure 8.5 presents research on blue carbon using ^{210}Pb, including the type, number of documents by country, and documents categorized by subject and year. Blue carbon studies using ^{210}Pb have experienced rapid growth over the past two decades. This distribution reveals that research is concentrated in developed coastal countries with strong blue carbon programs, while numerous countries remain

Table 8.1 Main challenges related to the use of radioanalytical methods

Challenges	Description
Cost and Infrastructure	Nuclear methods (such as radiometric dating, utilizing Pb-210 and Cs-137) typically require specialized laboratories and equipment, as well as expensive and secure handling facilities, and the regular import of reagents
Safety and Regulatory Constraints	Utilizing radioactive materials requires regulations: training, safety protocols, licensing, and disposal of radioactive waste
Sample Size, Preservation and Contamination	Radioanalytical methods often require well-preserved and non-contaminated samples. Diagenesis, bioturbation, mixing of sediments, or post-depositional alteration may disrupt radioisotope distributions, confusing interpretation. Additionally, achieving the necessary sample mass and quality is not always feasible, particularly for older or low-carbon sediments
Temporal Resolution and Dating Limits	As some radioisotopes have limited usable time scales (e.g., ^{14}C, which has a usable time scale of thousands of years, but with uncertainties, and ^{210}Pb, which has a usable time scale of the past ~100–150 years), dating becomes less accurate, mainly for recent or ancient carbon
Isotope Mobility/ Post-Depositional Redistribution	Radioisotopes in sediments may move after deposition and be redistributed—by diffusion, chemical migration, or groundwater flow—which can potentially mask age horizons/layers or tracer signals
Temporal and Spatial Coverage Gaps	Due to the costs and related efforts involved, radioisotopic research studies are less common in certain geographic zones, which may lead to biased global datasets

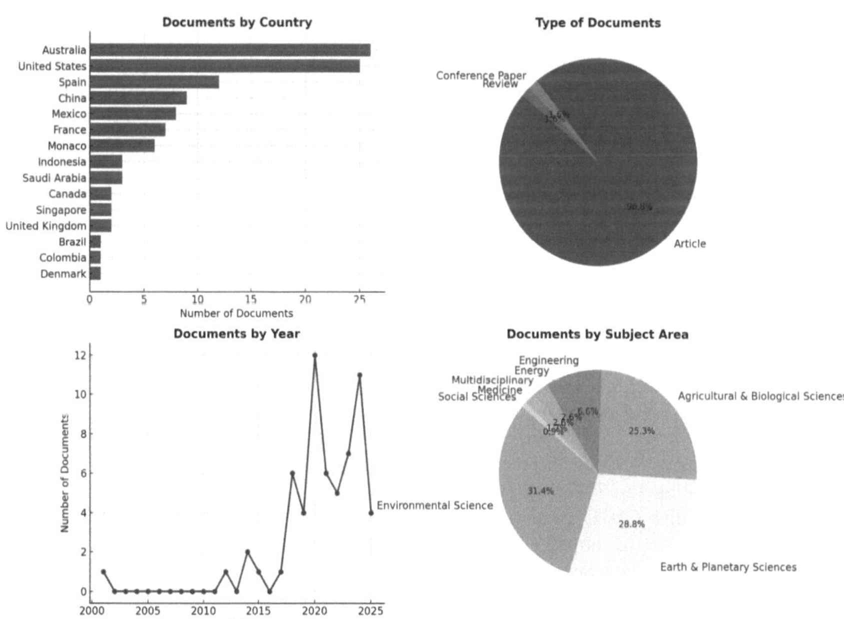

Fig. 8.5 Studies on blue carbon using ^{210}Pb radiometric dating distributed based on subject area, type of document, and presented by country and number of document by year. (*Data collected from Scopus Database (date: 18 September 2025)*)

underrepresented. Geographic gaps mainly exist in Africa, parts of Asia, and South America (except for Brazil), which can limit the overall representativeness of existing knowledge.

A dominance of original research papers was observed, suggesting that this field is still in its early stages, with a focus on research rather than synthesis or methodological reviews. While based on the subject area, the blue carbon field is highly interdisciplinary, primarily embedded in environmental and earth science research. This suggests that the application of Pb-210 is mainly used in sediment to reconstruct history, sedimentation rates, and carbon burial rates, providing needed insights for climate change mitigation strategies.

8.6 Conclusion

The analysis of nuclear and isotopic approaches employed in the field of blue carbon highlighted the role these techniques play in bridging gaps related to blue carbon. While there is a lack of data related to carbon sequestration and its long-term storage within blue carbon ecosystems, nuclear approaches have provided essential data to help assess these ecosystem capacities, thereby aiding in restoration and management programs for these natural ecosystems. Nevertheless, these approaches face several challenges, with cost, infrastructure demands, and accessibility remaining the primary limitations, resulting in geographical coverage gaps that necessitate expanding global coverage to evaluate blue carbon dynamics fully.

Acknowledgments This work was carried out within the framework of the Regional Project (AFRA) RAF7020. Project title: "Establishing Regional Capacities to Assess the Importance of Carbon Sequestration in Aquatic Systems for Climate Change Mitigation, Environmental Conservation and Economic Purposes (AFRA)". International Atomic Energy Agency (IAEA), 2022–2025.

References

Ahmed N, Glaser M (2016) Coastal aquaculture, mangrove deforestation and blue carbon emissions: is REDD+ a solution? Mar Policy 66:58–66. https://doi.org/10.1016/j.marpol.2016.01.011

Arias-Ortiz A, Masqué P, Garcia-Orellana J, Serrano O, Mazarrasa I, Marbà N, Duarte CM (2018) Reviews and syntheses: ^{210}Pb-derived sediment and carbon accumulation rates in vegetated coastal ecosystems—setting the record straight. Biogeosciences 15(22):6791–6818. https://doi.org/10.5194/bg-15-6791-2018

Blue Carbon Initiative (2021) Blue carbon and nationally determined contributions, 2nd edn. Blue Carbon Initiative. https://bluecarbonpartnership.org/wp-content/uploads/2023/10/Blue-Carbon-NDC-Guidelines-Second-Edition-2023_spreads.pdf. Accessed 21 Sept 2025

Campbell AD, Fatoyinbo L, Goldberg L, Lagomasino D (2022) Global hotspots of salt marsh change and carbon emissions. Nature 612(7941):701–706. https://doi.org/10.1038/s41586-022-05532-6

Coplen TB (2011) Guidelines and recommended terms for expression of stable-isotope-ratio and gas-ratio measurement results. Rapid Commun Mass Spectrom 25(17):2538–2560. https://doi.org/10.1002/rcm.5129

Daba DE, Dejene T (2018) Blue carbon: climate change mitigation, ecosystem services and coastal sustainability. Ocean Coast Manage 161:64–72. https://doi.org/10.1016/j.ocecoaman.2018.04.024

Den Hartog C, Kuo J (2006) Taxonomy and biogeography of seagrasses. In: Larkum AWD, Orth RJ, Duarte CM (eds) Seagrasses: biology, ecology and conservation. Springer, Dordrecht, pp 1–23. https://doi.org/10.1007/978-1-4020-2983-7_1

Duarte CM et al (2018) Recarbonization of the biosphere. Front Ecol Environ 16(6):321–327. https://doi.org/10.1002/fee.1790

Finlay JC, Kendall C (2007) Stable isotope tracing of temporal and spatial variability in organic matter sources to aquatic ecosystems. Limnol Oceanogr 52(3):937–947. https://doi.org/10.4319/lo.2007.52.3.0937

Fourqurean JW et al (2012) Seagrass ecosystems as a globally significant carbon stock. Nat Geosci 5:505–509. https://doi.org/10.1038/ngeo1477

Garcia FC et al (2019) The role of microbial processing in blue carbon cycling. Biogeosciences 16:2343–2357. https://doi.org/10.5194/bg-16-2343-2019

Geraldi NR et al (2019) Blue carbon sequestration in coastal ecosystems. Front Mar Sci 6:267. https://doi.org/10.3389/fmars.2019.00267

Goulden ML, Munger JW, Fan SM, Daube BC, Wofsy SC (1996) Measurements of carbon sequestration by long-term eddy covariance: methods and a critical evaluation of accuracy. Glob Change Biol 2(3):169–182. https://doi.org/10.1111/j.1365-2486.1996.tb00070.x

Howard J et al (2014) Coastal blue carbon: methods for assessing carbon stocks and emissions factors in mangroves, tidal salt marshes, and seagrass meadows. Conservation International, Intergovernmental Oceanographic Commission of UNESCO, International Union for Conservation of Nature

IAEA (2021) What is blue carbon? International Atomic Energy Agency. https://www.iaea.org/newscenter/news/what-is-blue-carbon. Accessed 21 Sept 2024

James PL, Larkum AWD (1996) Photosynthesis and biomass production of seagrasses. Aquat Bot 55:47–67. https://doi.org/10.1016/S0304-3770(96)01072-4

Jones MC et al (2018) Climate change impacts on coastal carbon sinks. Nat Clim Chang 8:12–20. https://doi.org/10.1038/s41558-017-0010-1

Ke X et al (2020) Carbon sequestration and storage in mangroves of China. Ecosyst Health Sustain 6(1):1740161. https://doi.org/10.1080/20964129.2020.1740161

Liu W et al (2023) Nuclear approaches in blue carbon science. J Environ Radioact 260:107071. https://doi.org/10.1016/j.jenvrad.2022.107071

Lovelock CE, Duarte CM (2019) Dimensions of blue carbon and emerging perspectives. Biol Lett 15(3):20180781. https://doi.org/10.1098/rsbl.2018.0781

Martin TR (2022) Coupling ^{210}Pb and ^{14}C to constrain carbon burial efficiency of blue carbon ecosystems. Master's thesis, University of South Florida

Mejjad N, Laissaoui A, El-Hammoumi O, Benmansour M, Benbrahim S, Bounouira H et al (2016) Sediment geochronology and geochemical behavior of major and rare earth elements in the Oualidia Lagoon in the western Morocco. J Radioanal Nucl Chem 309(3):1133–1143

Mejjad N, Laissaoui A, Ziad N, Ait Bouh H, El Hammoumi O, Benkdad A, Fekri A (2018) Evidence for a recent increase in delivery of atmospheric ^{210}Pb to Oualidia lagoon, coastal Morocco. Environ Monit Assess 190(11):642. https://doi.org/10.1007/s10661-018-6986-1

Mejjad N, Laissaoui A, Benmhammed A, Fekri A, El Hammoumi O, Benkdad A, Chakir EM (2022) Potential ecological risk assessment of rare earth elements in sediments cores from the Oualidia lagoon, Morocco. Soil Sediment Contam 31(8):941–958. https://doi.org/10.1080/15320383.2022.2067621

Mejjad N, Laissaoui A, El Hammoumi O, Fekri A, Amsil H (2023) Geochemical characterization of rare earth elements in sediment profiles from the Oualidia lagoon (Morocco). Acta Geochim 42(6):1051–1064. https://doi.org/10.1007/s11631-023-00643-1

Mejjad N, Safhi AEM, Laissaoui A (2025) Insightful analytical review of potential impacts of microplastic pollution on coastal and marine ecosystem services. J Hazard Mater Adv 17:100578. https://doi.org/10.1016/j.hazadv.2025.100578

Morgera E, Lennan M (2024) Marine biodiversity and climate governance. Cambridge University Press

Murray BC, Pendleton L, Jenkins WA, Sifleet S (2011) Green payments for blue carbon: economic incentives for protecting threatened coastal habitats. Nicholas Institute for Environmental Policy Solutions, Duke University

Nellemann C et al (2009) Blue carbon. A rapid response assessment. United Nations Environment Programme, GRID-Arendal

Petrokofsky G, Kanowski P, Brown ND (2012) Strengthening the impact of the forestry research: how can knowledge exchange enhance relevance and uptake? Forestry 85(4):491–500. https://doi.org/10.1093/forestry/cps046

Qureshi ME, Harrison SR, Wegener MK (2012) Estimating the impacts of climate change on net carbon emissions from Australian agriculture. Ecol Econ 81:37–45. https://doi.org/10.1016/j.ecolecon.2012.07.012

Samper-Villarreal J et al (2020) Carbon dynamics in Costa Rican seagrass beds. Estuar Coast Shelf Sci 236:106616. https://doi.org/10.1016/j.ecss.2020.106616

Schirone A et al (2022) Blue carbon ecosystems and radiometric dating. J Environ Radioact 255:106972. https://doi.org/10.1016/j.jenvrad.2022.106972

Schrag DP (2009) Storage of carbon dioxide in offshore sediments. Science 325(5948):1658–1659. https://doi.org/10.1126/science.1179510

Serrano O et al (2023) The role of blue carbon in climate change mitigation. Glob Change Biol 29(5):1178–1194. https://doi.org/10.1111/gcb.16530

Siikamäki J, Sanchirico JN, Jardine SL (2012a) Global economic potential for reducing carbon dioxide emissions from mangrove loss. Proc Natl Acad Sci USA 109(36):14369–14374. https://doi.org/10.1073/pnas.1200519109

Siikamäki J et al (2012b) Blue carbon: global options for reducing emissions. Environ Sci Pol 14(3):237–246. https://doi.org/10.1016/j.envsci.2011.12.010

Singh GG et al (2022) Carbon sequestration potential of coastal wetlands. Sci Total Environ 806:150548. https://doi.org/10.1016/j.scitotenv.2021.150548

Smith P et al (2020) Land–climate interactions. Nat Clim Chang 10(5):429–435. https://doi.org/10.1038/s41558-020-0791-5

Stevenson JC (1988) Comparative ecology of submersed grass beds in freshwater, estuarine, and marine environments. Limnol Oceanogr 33(4, part 2):867–893. https://doi.org/10.4319/lo.1988.33.4part2.0867

Tang D, Liu X, Xia Z, Hou J, Yang X, Li P, Yuan X (2023) Sources of organic matter and carbon stocks in two mangrove sediment cores and surface sediment samples from Qinglan Bay, China. Sci Total Environ 893:164897. https://doi.org/10.1016/j.scitotenv.2023.164897

The Information Architects of Encyclopaedia Britannica (2025) Kyoto Protocol. Encyclopedia Britannica. https://www.britannica.com/facts/Kyoto-Protocol

UNEP (2014) The importance of mangroves to people: a call to action. United Nations Environment Programme

Worthington T et al (2024) A global carbon stock assessment of coastal ecosystems. Nat Sustain 7:112–123. https://doi.org/10.1038/s41893-024-01234-5

Zieman JC (1982) The ecology of the seagrasses of South Florida: a community profile. U.S. Fish & Wildlife Service, Washington, DC

Zieman JC, Zieman RT (1989) The ecology of the seagrass meadows of the west coast of Florida: a community profile. U.S. Fish & Wildlife Service, Washington, DC

Zigah PK, Minor EC, Werne JP, Schwarz J (2012) Sources and cycling of dissolved organic matter in northern temperate lakes: insights from stable isotope analyses. Mar Chem 128–129:13–26. https://doi.org/10.1016/j.marchem.2011.11.001

Zigah PK et al (2017) Deciphering sources and cycling of dissolved organic matter in coastal waters using isotopes. Limnol Oceanogr 62(3):1014–1033. https://doi.org/10.1002/lno.10481

Chapter 9
Application of Nuclear Techniques and AI Curbing Seawater Pollution Addressing SDG 14 (Life Below Water)

Bhupinder Singh, Saurabh Chandra, and Ashima Jain

Abstract Enhancing nuclear technology into battling with seawater pollution is one of the most complex challenges; however, it gives a particularly robust aspect to artificial intelligence for achieving SDG 14 (Life Below Water). For example, this kind of tracking radioisotope makes it possible for scientists to determine, trace and quantify contamination, sourcing pollution, and monitoring global changes in the oceans very accurately. These processes are very important in depositing heavy metals, cleaning up oil spills and potentially radioactive activity in marine systems. In addition to nuclear approaches, there is an AI approach in working with massive datasets derived from satellite images, and remote sensing technology, which can model pollution patterns and identify significant hotspots. Machine learning sports tools that would classify patterns in pollution using historic and up-to-date information and encourage the prediction of strategic models on interventions regarding pollution for a better treatment of prevention and the distribution of resources for optimum efficacy. This is achieved through the use of AI-managed robots and drones in water to survey and clean up the pollution as well as to carry out environmental assessments by removing physical garbage.

Keywords Seawater pollution · Nuclear techniques · Artificial intelligence (AI) · Sustainable Development Goal 14 · Marine ecosystem protection

B. Singh (✉)
Sharda School of Law, Sharda University, Greater Noida, Uttar Pradesh, India

S. Chandra
School of Law, Bennett University, Greater Noida, Uttar Pradesh, India

A. Jain
School of Law, Manipal University Jaipur, Jaipur, Rajasthan, India

N. Mejjad (ed.), *Nuclear-Based Approaches application for marine pollution assessment and monitoring*, Environmental Science and Engineering,
https://doi.org/10.1007/978-3-032-12197-4_9

9.1 Introduction

Industrialization was able to bring about increased human welfare, but, at the same time, it also raised several environmental issues. This was brought about by industrial pollution and soaring human population tags on the limited renewable natural resources that led to severe environmental degradation. Large communities of people stress available resources but, through strong industrial emissions, also pollute air, water, and food, harmful to both aquatic and maritime habitats. This, in the long term, causes anthropogenic climate change, habitat loss, and would also bring in greatly altered environmental norms that might not be in favour of the Earth itself. The increasing scope of these shifts on health would necessarily enhance the need for better environmental surveillance of air, water, and soil pollution (Liu et al. 2024).

This type of recording is not only required to minimise environmental harm, but also to promote sustainable development and ensure that resources are available for the generations to come. On the other hand, with the fusion of nuclear studies and artificial intelligence, what becomes apparent are the vast interrelations among pollutants and marine ecosystems, opening the path to policy-making, which must be based on solid scientific evidence. This technology also helps evaluate how human actions such as overfishing, releasing industrial effluents, and plastic garbage are affecting ocean health. The implementation of these innovations will require collaboration between governments, researchers, and the industrial technocrats at the global level. Much of the promise may be seen in the areas of marine conservation, ocean air pollution reduction, and the very strong possibility of long-term restoration of ecological life, achieving, thus, a sustainable future for aquatic environments.

At present, technologies are being developed in exploding speeds in the world and changing many fields like speech recognition, language translation, biotechnology, quantum technology, virtual reality, robotics, machine vision and navigation. Advances in this field are dramatically changing how we work and how we think, at the same time as new solutions to a wide range of problems are emerging. Of these emerging technologies, Artificial Intelligence (AI) in conjunction with Machine Learning (ML) stands out as a promising technology to address environmental problems.

AI are the building blocks with which to create the autonomous devices, and machine learning enable these systems to use historical data to make predictions and reasoned decisions. In the framework of safeguarding the oceans, it is anticipated that machine learning could play a critical role in understanding oceanic ecosystems, monitoring shipping traffic, tracking marine debris, aiding ocean mining, combating illegal, unreported, and unregulated (IUU) fishing, and preventing coral-bleaching or marine disease outbreaks. Discoveries of such nature can lead to the development of applications or devices that can battle marine pollution on an effective note. By harnessing the combined capabilities of AI and machine learning, we could enhance protection of marine ecosystems, create pollution reduction strategies, and enforce sustainable development efforts for ocean resources.

The article attempts to put forth an analysis of the possibilities of AI combating marine pollution along with its concerns from the sustainability viewpoint. In this regard, it examines a variety of AI use cases in marine pollutants management and evaluates their scalability. While incomplete, they attempt to stimulate the development of larger and diversified solutions in full acknowledgment of the fact that AI progresses while becoming increasingly intelligent over the years (Crépin 2018).

9.1.1 Background of the Chapter

The Industrial Revolution brought enormous growth in per capita Gross Domestic Product and the population of the world. These advancements accelerated urbanization which created social problems including; congested, unhealthy and dilapidated housing, air and water pollution, and gigantic waste production which caused illnesses and polluted water. Indeed, some of the reforms made after the Black Death enhanced cleanliness and people's quality of life, however, it was not long before the pollution of the environment by industries through factors such as air and water pollution was not considered as it is today due to probably a lack of understanding about the issues between the medieval period and modernity. These issues deepened over the years which led to pollution being one of the many significant issues recorded (Agarwala 2020).

This meant that as the word population increased, the loss of land source lead to the exploitation of seas for foods and minerals. This exploitation coupled with contamination resulting from physical activities which contributes over 80% of marine pollution has greatly worsened the marine environment, modern seas have become giant dust bins almost on the brink of collapsing. Unfortunately, evidence shows that approximately by 2050, the amount of plastic in the ocean may exceed the amount of fish.

Besides these challenges, the wrong use of fossil fuels industrialization, and cement production accelerated climate change leading to drastic changes in ocean systems including the rise in sea level temperatures as well as the loss of the natural ability to absorb carbon dioxide. Realising the urgency of the rights and wrongs of the destruction of the environment and human effect, the United Nations has dedicated 7 of its 17 Sustainable Development Goals to these important causes.

As people are aware because of debates, its impact as a purveyor of pollution in terms of energy it consumes as well as resources that it affronts is undeniable, the same advancement is able to address environmental problems effectively. The new fresh ways of identifying and monitoring pollution cover laser technologies, chemical processes, nanotechnologies, and sensors. Also, technology helps to create sustainable material, new solutions, and a better understanding of the ways nature works. While the technological progress during the first industrial revolution is often directly linked to the modern environmental issues, Fourth industrial revolution is already bringing such developments to combat the environmental harms and guarantee the longevity of life on the planet (Biermann et al. 2020).

AI is the greatest revolutions of the Fourth Industrial Revolution which has a huge interference in almost every aspects of our life. AI is a perfect tool for fighting destruction of the environment a threat to world sustainability since the technology has many applications and its uses are only increasing nowadays. However, potential challenges like as high costs on implementation and need of regulation often act as barriers to its adoption. AI means the design of smart computer systems that emulate people's thoughts and actions. Maximizing resource utilization, AI with ML-DL can solve most eco-complex procedures and predict resource sustainability. All these achievements are inversely linked to the attainment of Sustainable Development Goal 14 that focuses on conservation and sustainable use of the ocean's resources (European Parliament 2020).

9.1.2 Evolution of Detection Methods

AI is a perfect tool for fighting destruction of the environment a threat to world sustainability since the technology has many applications and its uses are only increasing nowadays. However, potential challenges like as high costs on implementation and need of regulation often act as barriers to its adoption. AI means the design of smart computer systems that emulate people's thoughts and actions. Maximizing resource utilization, AI with ML-DL can solve most eco-complex procedures and predict resource sustainability. All these achievements are inversely linked to the attainment of Sustainable Development Goal 14 that focuses on conservation and sustainable use of the ocean's resources. Lacking the marked about the control, turning toward the monitoring and data collecting MOS, that enable the analysis as well as provided support in decision-making (Food and Agricultural Organisation (FAO) 2020).

There seems to be no better tool, at least from a bird's eye view, for monitoring large swaths of the ocean. Aeroplanes were used for this purpose but they were not ideal since they could only take data under favorable conditions and interval was long, data for specific point was not continuous and long term. To overcome these problems satellite technology was made available for land use from 1972 and later same was done for marine environment in late 1970s. This was the beginning of satellite navigation technology which enabled to identify, track, evaluate and observe maritime pollution.

With the integration of different sensors, which serve different purposes, remote sensing has evolved into an elastic tool backed by several carriers. However, each sensor type has its advantages and operating limitations and requires peculiar techniques to process images and receive useful data. This means that the world of industry sensing has tax on enhancing data processing methods. Various physical and electronic instruments over time have been used to measure important ocean variables including temperature, conductivity, pH, salinity, dissolved oxygen, chlorophyll fluorescence, viscosity, color dissolved organic matter (CDOM) and marine pollutants. These systems comprise orbiting satellites, research vessels, ocean observatories and marine robotics such surface vehicles, wave piercing catamarans and submersibles.

Modern software and current computers' computation capabilities have significantly expanded the array of monitoring approaches and indicators. These technical advancements have led to an exponential increase in the amount of data collected making ocean sensing a big data field. At the same time, the field of Machine Learning (ML) emerged to deal with and analyze this amount of information. Most big data technologies have now become popular given that; machine learning (ML) is a technique that deploys algorithms to filter data in an attempt to recognize certain patterns and as well to guess based on authentic info. The evolution to Deep Learning (DL) the use of programmable neural networks (NN) enhances its capability of accepting big data (United States International Trade Commission 2021).

As the name literally suggests, therefore, Machine Learning is the process of training machines to recognize and learn patterns. Information of varying types is employed to teach machines to recognize, predict and evaluate specific goals. In particular, information collected by airborne sensors is often in the form of light signals. Water and pollutants have different infrared characteristics that mean, they retain light differently; therefore, objects on the ocean's surface can be easily distinguished. These above differences the machines are enabled to sort and classify floating objects of water such as trash, oil, algae, seaweed, seafoam and floatwood.

It also turns out that machine learning algorithms are inadequate for extracting locally produced signals that can be found in just a negligible fraction of satellite images, and it also does not work for the long series data. The gap focuses on the importance of new data mining algorithms, and this can create or simply the way toward better technologies of deep learning. While using satellite to map the surface pollutant, subsurface pollutants requires special marine systems referred to as autonomous underwater vehicle-AUVs and remotely operated vehicles-ROVs to get essential data for brands that employ machine learning applications. ML and AI potential for recognizing maritime ecological pollution causes will be explored by examining various use cases that demonstrate growing opportunities and risks. All these examples help extend the list of the potential many solutions that AI may introduce, and the ability of those solutions to achieve far greater things (Choney 2020).

9.2 Sustainable Use and Conservation of Oceans, Seas and Marine Resources

The central legal framework for most sea operations is the Convention of the United Nations of 1982 on the Law of the Sea. UNCLOS addresses all aspects of maritime issues, without prejudice to protection and utilization of the marine species. This global legal system is directly linked with the SDG 14 which aims at conserving life beneath water (Costello 2020). The seas provide a twofold obstacle as while on the one hand there are objectives of conserving and preserving the water resources and the biological resources that dwell in water, on the other end there are goals of satisfying the needs of people who depend on water endowments as their resources. Article 61 of UNCLOS recognises the coastal States' right to limit the catch of living resources within their respective Economic Exclusive Zone. The intent of the coastal state shall ensure, to the absolute best of the knowledge which is scientifically available, that the exploitation of biological resources in the EEZ does not compromise its sustainability. Similarly, successful conservation and management require cooperation between the coastal states and suitable worldwide organisations at the sub regional, regional and global levels.

These UNCLOS commitments provide the foundation for SDG 14 whose objectives are to "Conserve and sustainably use the oceans, seas and marine resources for development." The term 'sustainability, therefore, has a double meaning that refers not only to sustained output and 'sustainable development' but also 'ecological sustainability'. The latter lays stress on both man's economic endeavour and nature, stressing rational interconnectedness of both aspects in space. This broaden perspective of sustainability may provide some resolution to the apparent paradox in the SDG 14 and conformity between the conservation of the sea's resources while ensuring sustainable utilization of the latter. It revolves around the need to strike a dynamic balance between people's utilization of resources to feed their economy and the needs of the environment (Goettsche-Wanli 2018).

The second target 14.2 states, "sustainably manage and protect coastal and marine ecosystems in order that they can better cope with significant negative impacts, prevent two thirds of degradation, and restore to achieve healthy and productive oceans". Quite simply, the performance of this target is measured by the "number nations adopting eco-approaches to managing marine areas." EBMs stress connectivity among species, their habitats, chemical, and physical factors in an ecosystem; resiliency of ecological systems, environment, and general well-being of ecosystem. These methods aim at reducing catastrophes while, at the same time, enabling individuals to continue to be in a position to benefit from services in nature enhancing the socio-ecological resilience. This idea also holds true for the management methodologies where ecosystem methods consist of management techniques for economic as well as social ecosystems, which are sustainable, societal, and economical with regards to the ethical development principles (Diaz 2020).

There is urgent need for intensified cooperation and synergy among nations in implementing diverse aspects of the UNCLOS as a foundation of SDG 14 and sustainable ocean management. The Division for Ocean Affairs and the Law of the Sea (DOALOS) of the United Nations General Assembly affiliate remains mandated to provide international assistance in this regard.

Article 61 has provided the meaning of "preservation" of living aquatic resources which implies that any coastal State shall set an appropriate catch limit to aquatic life in its special Economic Zone. Additionally, the states of the coastal states must adopt viable conservation and sustainable uses measures to avoid the negative effects of over harvesting on stock in the exclusive economic zone. These methods ought to: seek to conserve or rehabilitating species captured to a level that can yield the maximum sustainable output. This goal has to be achieved considering such factors as pressures from coastal fishing communities, and inexorable constraints of development in developing nations. In addition, the conservation strategy must address the fishing trends, relationships of species and minimum standard internationally at sub-regional, regional, and global levels (Fulton et al. 2018).

Coastal and marine area target is linked to the SDG 14.5 in which at least 10% of those areas should be conserved by the year 2020 through national and international laws with the help of the best available scientific evidence. The first measure for this objective includes percentage of marine regions protected by specific areas. Marine protected areas play a key function in mitigation of the decline of natural diversity and on fair use of marine resources. These are legally defined areas within or in proximity to the maritime environment experiencing higher levels of protection than their surroundings.

SDG 14.a focuses on the appropriateness of choice, improved and enhanced scientific throughput, enhanced research capability, and ocean-based technology dissemination. This method is indeed important in support of the development of the new generation of nations, SIDs and the least developed ones (Hafeez et al. 2018).

9.2.1 Use-Cases

For better comprehension of how longevity of the concept can aid in the conservation of the marine region we have the following points stating several working models of AI. Below are what can be perceived as application examples touching nearly all concerns as an evidence of the capacities of AI to tackle these challenges from different angles. To increase the level of detail, the use scenarios have been separated by certain problem fields. It is important to stress that here we outlined only several examples and it was discussed what roles the AI can take to protect the aquatic ecosystems, but it does not list all the opportunities of its application (Harris 2018).

1. **Oil spills**

   Oil spills may be in the form of natural disasters or because of human actions. Spills that are characterized as natural are normally very small in scale, contributing to only about 10% of all occurrences. The type of human triggered leakages may be intentional like the release of crude oil from the tanks of the ships, or they often occur as a result of horrible disasters at sea. Oil spills are a continuous and move along the stream of water currents, which means that monitoring requires both floating and stationary equipment. The fixed sensors in use include side-looking airborne radar, laser fluorosensors, as well as ultraviolet and infrared heating cameras and are particularly efficient where the area of the spillage is already known. The characteristic of oil spills as a process that may rapidly cover large areas, makes it impossible to identify the affected zones in real time, thus rendering the use of fixed sensors to monitor the process ineffective.

SAR has been proved to be an effective way in monitoring monitoring oil spills. It functions by measuring the unevenness of the sea appear, that is, thinning in oil-covered areas due to freeing surface tension. However, it was noted that seaweeds were equally capable of producing such a surface roughness that it is almost impossible to differentiate between the two. In most cases, human intervention is needed to differentiate between the natural features and the harshest environment caused by oil spills (Just 2020).

Due to the improvements in Artificial Intelligence (AI) this separation had become more precise and thus more effective in measuring the surface roughness in locations that are contaminated with oils. One advantage of SAR is that it is based on satellites, and therefore it can also study areas where people cannot reach in person. This equipment has to do a lot with detection of oil spills and is essential to save human and marine life. It also plays a significant function in averting interceptions of unlawful discharge, which contribute to ocean pollution (Lei et al. 2019).

One such system, which is already in operation today, is the Canadian Environmental Hazards Detection System (CEHDS). This system analyses photos from two satellites: Canadian Radar Satellite RADARSAT and European Remote Sensing Satellite-1 (ERS-1). It uses the 1-nearest-neighbor (1-NN) rule with a SHRINK learning method that helps it make decisions. In so doing, CEHDS can easily train itself with new data that come in, which means that it is highly flexible and will remain effective in oil spill monitoring always.

2. **Monitoring marine life**

   Machine instruction and neural networks greatly combined to augment the intelligence level required for face detection. Scholars have applied this technique with variations in the patches on the whale fins, flukes, and whale sharks. This capacity permits utilizing video and photo footage to track organisms in a given area and thereby minimize interference with the variety, in addition to keeping images for future research objectives. Popular tools for marine living assets identification and mapping include iNaturalist, which shares its data through the Global Biodiversity Information Facility, Ecotaxa, and a plankton identification card known as PIC. In addition, tools such as Linne Lens, a device that uses smartphone for analyzing a variety of marine species live and in the same process; FathomNet and Video and Image Analytics for Marine Environments (VIAME) are vital. Squidle+, BISQUE—Bio-Image Semantic Query User Environment for preservation and annotating marine images, biome.com for showing environment problems in Japan, and WildTrack for tracking animals on foot prints through images aid in supporting development in marine animals tracking (Martens et al. 2025).

3. **Monitoring coral health**

   Approximately 90% of the world's corals may die even if fossil fuel use is phased out according to scientists' findings. Such corals play an important role in the ocean's ecosystem, and their disappearance means death for many marine inhabitants. Hence, it is necessary to preserve and, if possible, rehabilitate them. In order to correct this, several measures have been undertaken with an aim of mapping and studying coral reefs so as not cause further deterioration. For instance, CoralNet is an AI that is in the public domain and works by using benthic photos to assess corals and rocky outcrops (Chenillat et al. 2021). This software operates as a joint data sharing center and also supports both fully automated and semiautomated picture annotation by the use of deep neural networks.

Moreover, it is noteworthy that, Intel, Accenture, and the Sulubaaï Environmental Foundation of Philippines created the CORail to protect coral reefs. This works employs artificial intelligence and video analytics, deploying submersible cameras and roving cameras throughout the process, making the use of costly long-term study unnecessary (Li et al. 2020). The obtained data can be useful for researchers trying to find techniques for protecting and preserving the marine environment. It is also applied in tracking movements of Tropical Fish in the water. Other programs that are working towards this are the Allen Coral Atlas, Shell Ocean Discovery XPrize and LarvalBot in the attempt to come up with techniques that can be used to reseed the delicate coral reefs.

4. **Underwater mapping**

The ocean deals with the weather but to understand how it really is and its valuable assets which are foods, metals and energy among others, we need to know more about the ocean and the seabed. Even while examining and imaging aerially, by means of drones, satellites, examinations and imaging vehicles, useful seabed maps have only been created out of less than 10% of the ocean floor and the 4000 m plus deep areas are virtually uncharted. Seabed monitoring and assessment are crucial to understanding ocean dynamics, seabed morphology, benthic processes, impacts of human activities and effects of global climate change on marine habitats. In order to do this, automation and new technologies including AI need to improve the processing of a large amount of data, and provide faster insights and assurance of sustainable human activities in other sensitive locations in the maritime environment.

To address these problems, several AI-enabled initiatives have been developed. For instance, Project Natick is an underwater data centre that is responsible for the global management of data and is equipped with an AI system that carries out a study on the potential environmental impact of the tender by studying aquatic life around it (Macaulay 2020). This technology is scalable, fully automated, and may be utilized for other applications, such as tracking environmental effects. Another project aims at enhancing undersea mapping for far better seabed characterization employing artificial intelligence to analyze sonar data. This system helps in determining the right locations where to install moni toring stations to obtain comprehensible seabed data. In addition, there are app solutions like SINAY and Data360 built to cover obvious knowledge gaps (Ukaogo et al. 2020). These app sort information from many sources including IoT sensors and marine sounds and through use of ML provide means of an immediate action while at the same time creating awareness on the deterioration of marine environment.

5. **Plastic/debris monitoring and removal**

   Some pollutants dispense from land to ocean because these two regions are connected. It will hardly be a shock to anyone that what is done on land has consequences for the state of the seas. Already water pollution originating from the land is highly responsible for ocean pollution, and here the most dominant source is plastics. These contaminants are often discharged deliberately or else brought into the ocean by water and wind currents. The consequence is extensive impacts on marine and aquatic environments and the organisms inhabiting these systems. Plastics which are in the water are of different types having different sizes and estoys degrade for several years. What is more, unlike the floating trash, a large portion of this trash effectively submerges into the water (Xu et al. 2020). These particles continue to accumulate in vast regions called ocean gyres where currents result in immense regions of shed particles floating. These gyres include the Great Pacific Garbage Patch (GPGP), one of the worst environmental hazards in the globe stretching over an area of approximately 1.6 million square kilometres (617,763 square miles).

   Therefore, the scale of this environmental disaster requires relevant and efficient interventions. In response to this, two important techniques are being developed. The first is to take care of currents sea trashing while the second is even more ambitious and wants to prevent future trashing by identifying its sources. They both heavily depend on the implementation of artificial intelligence (AI) functionality enabling deploying real-time and independent processes that enhance the effectiveness of these campaigns. With the help of AI these technologies carry a rather interesting potential in addressing the adverted issue of ocean pollution which has now become a big challenge to the marine ecosystem and the health of oceans.

9.3 Protecting the Economy Surrounding Oceans

9.3.1 Sustainable Fisheries

Fisheries are protected and managed for sustainable use by the United Nations Convention on the Law of the Sea popularly known as UNCLOS. Since the under UNCLOS Article 58 inside of EEZs where three-quarter of the world's fisheries are located, coastal governments have both the right and the duty for the use of fisheries responsibly. UNCLOS regulations on the particular enumeration and preservation and handling of straddling and extremely migratory fish stock, which were enshrined in the 1995 United Nations Fish Stocks Agreement, more commonly referred to as the UNFSA, contains an affirmative provision in article 6. Article 6 of the 1995 Agreement for the Implementation of UNCLOS regulations on the preservation and handling of straddling and extremely migratory fish stocks (commonly known as the United Nations Fish Stocks Agreement or UNFSA) promotes a proactive approach to the preservation and supervisors, and exploiting of these fish stocks (Zeng and Wang 2020; IPCC 2019a).

The agreement also stresses the need of nations having agreed access to and delivering scientific data, as well as making certain the effectiveness of conservation and management action. In addition, a UN General Assembly resolution requires states to employ precautionary and ecosystem approaches when taking conservation and management actions. These measures should resolve issues examples of which are bycatch, pollution, overfishing, and important habitats preservation in line with the modern UN Food and Agriculture Organisation guidelines.

SDG 14.4 aims at improving the efficiency of fishing activities so as to eliminate overfishing, and other related immoral fishing practices within 10 years. By 2020, it sought to apply ecosystem approaches to fisheries to achieve the rebuilding of fish stocks to productive capacity to allow maximum sustainable yield that various biological characteristics of the species can offer. Monitoring and verification of the status of fish stocks are done by an indicator called 'Indicator 14.4.1—The proportion of fish stocks that are within biologically sustainable levels'. A fish stock is described to be physiologically sustainable if it is able to support or exceed the biomass level that can support the Maximum Sustainable Yield. The term "maximum sustainable yield" therefore relates to the largest quantity catch that maybe harvested from a stock inventories to nearby unchanging environmental conditions such as the ecosystem, water supply, the species variability, and any factors that touch on birth, growth, or death rates of the species (Zhang et al. 2023). Although it falls within the UNCLOS, UNFSA, and the FAO Code of Conduct for Responsible Fisheries (1995) guidelines, the emphasis of the evaluation of feasible fish stocks based on their ability to generate sustainable profits appears to be somewhat more consumption-oriented than conservation-oriented.

The FAO Port State Measures Agreement (PSMA), which was adopted in 2009, is the first international instrument in the world designed to combat IUU fishing. This is the first contractual provision of this nature that has included port-state measures in order to discourage ships from engaging IUI fishing activities. Specifically, no ports or new transshipment opportunities are to be provided to unauthorised or IUU vessels and consequently no catches from such vessels are to be released, meaning that nobody has rights to sell or import a piece of such misunderstood and condemned merchandise on the markets of the country and globally. The PSMA incorporates the notion of IUU fishing format given in FAO's 2001 IPO AIPDEIUFG. The treaty was signed in 2016 and can be currently observed with 68 States Parties. As defined by the FAO, IUU fishing is any fishing operation that transgresses fisheries law and policy in territorial sea, restricted fishing zones or high seas.

This is one of the most effective tactics that have been used to eliminated illicit fishing and this is through vessel flags. A state has to limit the ability of ships' re-flagging (as it is described in the Article 5) and enhance its control to ensure compliance with international standards of preservation and management of vessels. Compliance Agreement refers to the Agreement to Facilitate Adherence with International Conservation and Management Practices by Fishing Vessels in the

High Seas made in effort to enhance flag states' responsibilities and enforce the high seas compliance with international standards.

Moreover, the Agreement on the Conservation and Management of Straddling and Highly Migratory Fish Stocks under the UNFALACS commenced on 11 Dec 2001 to guarantee the consistent preservation and ethical usage of these fish inventory under the provision of the UNCLOS. The UN Fish Stocks Agreement explains the obligations of flag states regarding vessels' identification and documents, authorization procedures, monitoring and control, surveillance (MCS), and obedience.

9.4 Some Regional Organisations Are Supporting Various International Agreements that Can Help to Fight Against IUU Fishing

9.4.1 Subsidies Contributing to Illegal, Unreported and Unregulated Fishing

This paper tries to argue that subsidies in fishing industry play a large role in overcapacity at fishing, overfishing, and even in the cases of IUU fishing. Target 14.6 for the SDGs suggests the need to prohibit or immediately eliminate such subsidies and that poor and least developed countries must be given special and differential treatment in WTO talks on fisheries subsidies. Regarding SDG14.6, progress is evaluated in the degree of governments' adoption of international instruments intended to combat IUU fishing (FAO 2020). The United Nations Convention on the Law of the Sea, the United Nations Fish Stocks Agreement, the International Plan of Action to Prevent, Deter, and Eliminate IUU Fishing, the FAO Agreement on Port State Measures from 2009, the FAO Voluntary Guidelines for Flag State Performance, and the FAO Compliance Agreement (UNESCO 2021–2030). While SDG 14.6 addresses the removal of harmful subsidies, the indicator to measure this progress refers to wider adherence to many international treaties and thus may overlap with SDG 14.4.

9.5 The Enforcement of a 'Right to Healthy Oceans, Seas and Marine Resources'

SDG 14, dedicated to the sustainable management of oceans and marine ecosystems, has close links with the regulations of the United Nations Convention on the Law of the Sea (UNCLOS) and legal systems governing maritime domains. UNCLOS embodies several concepts inducing various state responsibilities, including that of the duty to prevent, mitigate, and regulate the harmful effects of marine pollution from any source, as expressed in Article 194(1). Nevertheless, there is great uncertainty as to what legal substance these frameworks may offer. For instance, the existence of a ‘right to healthy oceans, seas, and marine resources’ beyond the 200-mile limit of the continental shelf is one enormous area of grayness left to be made clear; essentially fleshing out the way a government should act in such regions and how such freedoms can be maintained.

Controversy will become even more outrageous, particularly in regions like the Arctic, Antarctic, and the East and South China Seas where rival territorial claims between different states will undermine the efforts of international organizations such as the International Tribunal for the Law of the Sea (ITLOS) and Commission on the Limits of the Continental Shelf (CLCS). The unresolved conflicts in these areas bring to light the limits of enforcing ITLOS decisions, including the absence of adequate autonomous powers for enforcement and the relatively limited suggestions of the CLCS. Even monitoring systems exist, compliance mainly depends on states’ willingness to adhere to the judgments and guidelines.

Reference to the need to protect the maritime environment beyond the national jurisdiction confirmed that each State Party must have a right to bring a complaint. In fact, the Chamber thereby acknowledged the existence of the actio popularis for damage to worldwide commons. Recognition of Article 48 of the ILC Articles on State Responsibility amounts to the Chamber’s being the first global objector to recognize its significance (IPCC 2019b). With the Seabed Mining Advisory Opinion admitting a legal basis of Article 48, this will increase its standing in international law. Given, this elaboration holds a big promise towards understanding how transnational courts could enforce sustainability responsibilities and thereby achieve the stated aims of SDG 14.

9.5.1 How Nuclear Techniques Can Help Tackle Marine Environmental Challenges

Nuclear technology is a basic response to challenges raised by climate change and oceans modifications. Nuclear probes and isotopes are useful tools for evaluating marine ecosystems’ response to acidification, as well as other stresses. This method will help pinpoint the sources of the pollution introduced into the seas, providing critical information to scientists and policymakers that can be used to make well-informed decisions regarding the protection of threatened ecosystems.

There appears to be an ever-increasing concern about the mounting threat posed by harmful algal blooms (HABs)—a natural phenomenon of marine habitats posing threats to human health and ecosystems, especially in coastal areas. Researchers focus recently on the implications of HABs potentially worsened by climate change and human activities. The IAEA Environmental Laboratory has been at the forefront of promoting the influence of a nuclear technology, known as Receptor Binding Assay (RBA). This sensitive technique can be applied to early identification and continuous monitoring of biotoxins produced by harmful algal blooms

(HABs). The successful applications are reported from Chile, El Salvador, Colombia, and Cuba. Carlos Alonso Hernandez, a Researcher at the IAEA Radioecology Laboratory, was therefore interested in confirming early detection limiting the impact of harmful algal blooms (HABs). He pointed out the importance of nuclear methods in rapidly detecting biotoxins in seafood or surrounding areas (Lu et al. 2024). This allows for better outbreak identification, protecting the food chain and shortening the time for fishing site closures.

9.6 The Way Ahead

Artificial intelligence has established significant advancements for maritime environmental protection by providing potential remedies to human-induced damages. While these use cases are aimed at commercially available products, many other projects are still being developed or proposed, such as the Integrated Marine Debris Observing System. With such state-of-the-art technology as AI now becoming more widely available, many organizations have tools and services to support efficient data acquisition, real-time analysis, and data processing. These advancements provide reports integrated on pollution response, debris removal, climate change impact evaluation, conservation activities, and the mapping of marine habitat.

With all these rapid advances in the field of AI and the growing number of organizations competing in the area of AI solutions for environmental monitoring such as Google, Stem, Hazama Ando, IBM, Nnergix Environmental Protection SL, Verdigris Technologies, Siemens, Honeywell International, Raycatch, HST Solar, and Upside Energy. As this happens, better and more accurate predictions regarding the next environmental monitoring can be made, allowing better decisions to be made regarding its sustainable management.

However, it is important to note that while AI is very useful for prediction and analysis of data, it delivers only information and predictions on which actions must be finally decided by a human. Our responsibility, though, is a must if salvation is to come to this world in the face of now-existing outstanding environmental cataclysms; if we fail to act now, this will lead to considerable environmental damage, leaving behind a nothingness for our children to inherit. “A Planet B” is not available; thus, every action must be taken as reasonable expediency to ensure and safeguard Earth’s continued existence and health. In this regard, artificial intelligence (AI) becomes a handy device worthy, to intervene promptly and guide remedial efforts before it is too late.

The availability of reliable baseline data is crucial to attaining sustainability. Therefore, efforts should focus on actively collecting or passively monitoring comprehensive data. Sufficient and credible data is needed to inform the design and efficient implementation of ML algorithms that will give a profoundly transformative insight into sustainable management and use of natural resources (Xixi et al. 2022). Information generated through AI solutions will have a key role in improving decision-making and policy development, as well as in supporting more effective environmental governance and protection actions.

9.7 Conclusion

With the scientific community increasingly conscious of monitoring different variables in oceans, for instance, climate change, weather patterns, natural catastrophes, and ocean resources, human interaction with marine environments has grown tremendously. Mechanisms of remote sensing have mushroomed, thus providing an avalanche of data (Kennish 2017). This quantum of data has built the need to have a lot of novel tools and methods of handling and analyzing them. In answer to this problem, AI has been used in many ways to organize and make sense of oceanographic data.

AI methods have been far more beneficial in environmental studies. At the heart of growth is the ability of AI to churn through and process tons of data at a fast pace, with insights and inferences that sometimes elude human researchers' minds. AI not only promotes data interchange and collaborative analysis but also strengthens joint work among researches. This shared nature of the method allows for the identification of patterns and changes that would otherwise go underreported, with minor environmental changes frequently gliding under the radar of human notice. Such unperceived alterations might herald dire consequences for the environment over the long haul, but they can be snagged by AI before they spiral into major troubles.

The role of AI in marine conservation through sustainability initiatives is becoming more apparent. With further development, the commercialization of AI technologies will introduce competition in the market, and data would be cheaper and available to a wider range of interested parties. ML and DL have become indispensable for analyzing data collected by remote sensing systems. Various study cases have illustrated the contributions AI made to marine conservation, although they should not always be considered conclusive. On the contrary, they can actually be considered as milestones on the way toward more holistic and diversified methodologies. By being its fairly new field, AI has so enlarged its scope for envisaging doing some environmental issues with greater intelligence and efficiency. While it is promising, the current application of AI in oceanic data analysis is still an emerging field, and it has enormous potential to drive innovative solutions to the preservation and sustainability of the marine environment. With the continued development of AI technologies, more powerful weapons for the researchers will arise for a deeper understanding of and solutions to the integrated complexities of the ocean environment.

References

Agarwala N (2020) Monitoring the ocean environment using robotic systems: advancements, trends, and challenges. Mar Technol Soc J 54(5):42–60. https://doi.org/10.4031/MTSJ.54.5.7

Biermann L, Clewley D, Martinez-Vicente V, Topouzelis K (2020) Finding plastic patches in coastal waters using optical satellite data. Sci Rep 10:5364. https://doi.org/10.1038/s41598-020-62298-z

Chenillat et al (2021) Fate of floating plastic debris released along the coasts in a global ocean model, 165. http://www.unesco.org/new/en/natural-sciences/ioc-oceans/focus-areas/rio-20-ocean/blueprint-for-the-future-we-want/marine-pollution/facts-and-figures-on-marine-pollution/; https://www.sciencedirect.com/topics/earth-and-planetary-sciences/marine-pollution. Accessed 12 Jan 25

Choney S (2020) Microsoft Hackathon leads to AI and sustainability collaboration to rid plastic from rivers and the ocean. Microsoft. https://news.microsoft.com/features/microsoft-hackathon-leads-to-ai-and-sustainability-collaboration-to-rid-plastic-from-rivers-and-the-ocean

Costello M (2020) AI is a promising new tool for monitoring marine biodiversity. University of Auckland. https://www.oceansofbiodiversity.auckland.ac.nz/2020/06/22/ai-is-a-promising-new-tool-for-monitoring-marine-biodiversity

Crépin A-S (2018) A holistic approach to Arctic management. Scenarios for a warmer Arctic, Stockholm, Sweden. https://www.kva.se/en/kalendarium/scenarios-for-a-warmer-arctic

Diaz J (2020) World Ocean Day: some examples of AI use in the maritime domain. ActuIA. https://www.actuia.com/english/world-ocean-day-some-examples-of-ai-use-in-the-maritime-domain

European Parliament (2020) The environmental impacts of plastics and micro-plastics use, waste and pollution: EU and national measures, 29

FAO (2020) The state of the world fisheries and aquaculture. Sustainability in action. FAO, p 133

Food and Agricultural Organisation (FAO) (2020) The state of the world fisheries and aquaculture. Sustainability in action. FAO, p 94

Fulton M, Jungseok H, Islam MJ, Sattar J (2018) Robotic detection of marine litter using deep visual detection models, computer science and engineering. In: Proceedings of the international conference on robotics and automation, Montreal, Canada. https://doi.org/10.1109/ICRA.2019.8793975

Goettsche-Wanli (2018) The role of the United Nations, including its Secretariat in Global Ocean Governance. In: Attard, Ong, Kritsiotis (eds) The IMLI treatise on global ocean governance: Volume I: UN and global ocean governance, p 13

Hafeez S, Wong MS, Abbas S, Kwok CYT, Nichol J, Lee KH, Tang D, Pun L (2018) Detection and monitoring of marine pollution using remote sensing technologies. In: Fouzia HB (ed) Monitoring of marine pollution. IntechOpen. https://doi.org/10.5772/intechopen.81657

Harris B (2018) The latest weapons in the fight against ocean plastic? Drones and an algorithm. World Economic Forum. https://www.weforum.org/agenda/2018/06/this-ai-is-learning-to-recognize-ocean-plastic-using-drone-photos

IPCC (2019a) Summary for policymakers. IPCC special report on the ocean and cryosphere in a changing climate, 5

IPCC (2019b) Summary for policymakers. In: Pörtner et al (eds) IPCC special report on the ocean and cryosphere in a changing climate. IPCC. https://www.ipcc.ch/site/assets/uploads/sites/3/2019/11/03_SROCC_SPM_FINAL.pdf. Accessed 12 Jan 2025

Just V (2020) Using AI to map marine environments. Phys.Org. https://phys.org/news/2020-05-ai-marine-environments.html

Kennish MJ (2017) Practical handbook of estuarine and marine pollution. CRC Press

Lei M, Yu L, Xueliang Z, Yuanxin Y, Gaofei Y, Brian AJ (2019) Deep learning in remote sensing applications: a meta-analysis and review. J Photogramm Remote Sens 152:166–177. https://doi.org/10.1016/j.isprsjprs.2019.04.015

Li X, Bin L, Gang Z, Yibin R, Shuangshang Z, Yingjie L, Le G, Yuhai L, Bin Z, Fan W (2020) Deep-learning-based information mining from ocean remote-sensing imagery. Natl Sci Rev 7(10):1584–1605. https://doi.org/10.1093/nsr/nwaa047

Liu X, Cao S, Liu Y (2024) Legal challenges of regulating cross-border nuclear pollution from ship ballast water under the background of Fukushima wastewater release and China's response. Mar Policy 167:106288

Lu L, Wu Y, Zhang L, Shi J (2024) Psychoeducational research on the wisdom of harmonious co-existence in marine ecological protection: an example of scientific education on Japanese release of radioactive wastewater. Ocean Coast Manage 248:106979

Macaulay T (2020) AI detects plastics in the oceans by analyzing satellite images. The Next Web. https://thenextweb.com/neural/2020/04/23/ai-detects-plastics-in-the-oceans-by-analyzing-satellite-images/

Martens, Ellmers, Pokorny on behalf of Global Policy Watch (2025) COVID-19 and the SDGs. The impact of the coronavirus pandemic on the global sustainability agenda, 6f. https://www.ecowatch.com/coronavirus-plastic-waste-2645831072.html. Accessed 12 Jan 2025

Ukaogo POE, Ugochukwu O, Chibuzo V (2020) Environmental pollution: causes, effects, and the remedies. In: Chowdhary P, Raj A, Verma D, Akhter Y (eds) Microorganisms for sustainable environment and health. Elsevier, pp 419–429. https://doi.org/10.1016/B978-0-12-819001-2.00021-8

UNESCO (2021–2030) The science we need for the ocean we want United Nations decade of ocean science for sustainable development

United States International Trade Commission (2021) Seafood obtained via illegal, unreported, and unregulated fishing. U.S. imports and economic impact on U.S. Commercial Fisheries. Report of the United States International Trade Commission, p 11

Xixi Z, Tongkun Q, Yecheng W (2022) Optimal strategies for stakeholders of Fukushima nuclear waste water discharge in Japan. Mar Policy 135:104881

Xu J, Wang H, Cui C, Zhao B, Li B (2020) Oil spill monitoring of shipborne radar image features using SVM and local adaptive threshold. Algorithms 13:69. https://doi.org/10.3390/a13030069

Zeng K, Wang YA (2020) Deep convolutional neural network for oil spill detection from Spaceborne SAR images. Remote Sens 12(6):1015. https://doi.org/10.3390/rs12061015

Zhang D, Chen B, Hubacek K, Meng J, Sun M, Mao J, Jin M, de Almeida CM, Chiu AS, Yang L, Zhang L (2023) Potential impacts of Fukushima nuclear wastewater discharge on nutrient supply and greenhouse gas emissions of food systems. Resour Conserv Recycl 193:106985

Chapter 10
Synergistic Applications of Microbial Biocatalysis and Nuclear Technologies in Marine Pollution Management

Nisha Shankhwar

Abstract Marine ecosystems face mounting pressure from pollutants such as industrial discharge, agricultural runoff, plastics, and oil spills. These pollutants build up in ecosystems over time and become stronger as they move up the food chain. Detection and monitoring are the main challenges in marine pollution management. Conventional cleanup strategies like filtration, biodegradation, chemical treatments, recycling are not effective. Traditional pollution detection and monitoring in ecosystems are often slow, costly, and inaccurate due to their reliance on on-site labor, single sensors, and manual data validation. Nuclear techniques strengthen these processes by enabling precise detection and monitoring. Isotopic labeling, radiometric assays, and Nuclear Magnetic Resonance spectroscopy provide detailed spatial and temporal profiles of contaminant dynamics, allowing refinement of bioremediation strategies. Microorganisms employ enzymatic pathways including hydrolysis, oxidation, reduction, and dehalogenation to convert pollutants into less harmful intermediates. The integration of microbial bioremediation with nuclear techniques offers an effective approach to addressing hydrocarbon and radioactive contamination. This method utilizes microbial processes to degrade pollutants while employing nuclear tools to enhance monitoring and control. Studies involving radiolabeled substrates and isotopic fractionation shed light on microbial degradation pathways, aiding in the enhancement of bioprocesses. Regulatory and technical challenges persist; the convergence of microbial and nuclear technologies offers a scalable framework for addressing complex marine pollutants. This chapter presents an integrated strategy combining microbial biocatalysis and nuclear technologies to mitigate human impacts and support long-term ocean preservation.

Keywords Nuclear technologies · Microbial biocatalysis · Environmental remediation · Isotopic tracing · Sustainable marine management

N. Shankhwar (✉)
Special Centre for Nanoscience, Jawaharlal Nehru University, New Delhi, India
e-mail: nisha2023@jnu.ac.in

N. Mejjad (ed.), *Nuclear-Based Approaches application for marine pollution assessment and monitoring*, Environmental Science and Engineering,
https://doi.org/10.1007/978-3-032-12197-4_10

10.1 Introduction

Marine pollution has emerged as a critical global challenge with profound consequences for both ocean ecosystems and human societies. The marine environment is contaminated by a diverse range of pollutants, including heavy metals, plastics in micro and nano form, petroleum hydrocarbons and excess nutrients. Global plastic production continues to rise, with hundreds of millions of metric tons produced each year. Large amounts of this waste enter the ocean, where plastics break down into persistent microplastics that accumulate in water, sediments, and marine organisms. These materials can harm marine life by reducing growth, reproduction, and survival, and their effects move up the food chain, threatening entire ecosystems. Plastic pollution in the ocean has become a widespread environmental problem with serious consequences for both marine biodiversity and human communities that depend on ocean resources. Micropollutants are especially concerning because they stay in the ocean for a long time, build up in the bodies of marine animals, and cause lasting harm to ecosystems. As these pollutants move through the food chain, they can also affect the safety of seafood and the health of people who depend on it (Sinha 2024; Hermabessiere et al. 2017; Baek et al. 2022). Communities that rely on fishing and other ocean resources are especially vulnerable because pollutants build up as they move through the food chain, threatening both food security and people's livelihoods.

10.1.1 The Marine Pollution Challenge

It is proven that even relatively low concentrations of pollutants can exert serious effects on marine organisms and, by extension, the stability of entire ecosystems. Pharmaceuticals like antibiotics and synthetic hormones are concerning because they can disrupt the hormone systems of marine animals. This can affect their growth, reproduction, and survival, and even small amounts can build up in the food chain, harming entire ecosystems in long run. These biological disturbances propagate through food webs, altering species interactions and ecosystem dynamics. This adversity go beyond the ecosystems, since damage to ocean also puts communities and economies at risk, especially those that depend on ocean resources (Bilal et al. 2022). Pesticides such as organophosphates and carbamates don't just target pests; they also disrupt the nervous systems of many aquatic species. Meanwhile, microplastics, tiny plastic fragments, are often ingested by marine animals, mistaking them for food. Not only can they cause physical damage, but they also sneak toxic chemicals into the entire food chain (Gaw et al. 2014). Micropollutants have made their way even into the deep sea, showing up in waters as deep as 3000–10,000 m. It's pretty wild when you think about it: these tiny contaminants spreading so far and sticking around for so long, proving just how deeply they've embedded themselves in marine ecosystems (Spindola Vilela et al. 2022).

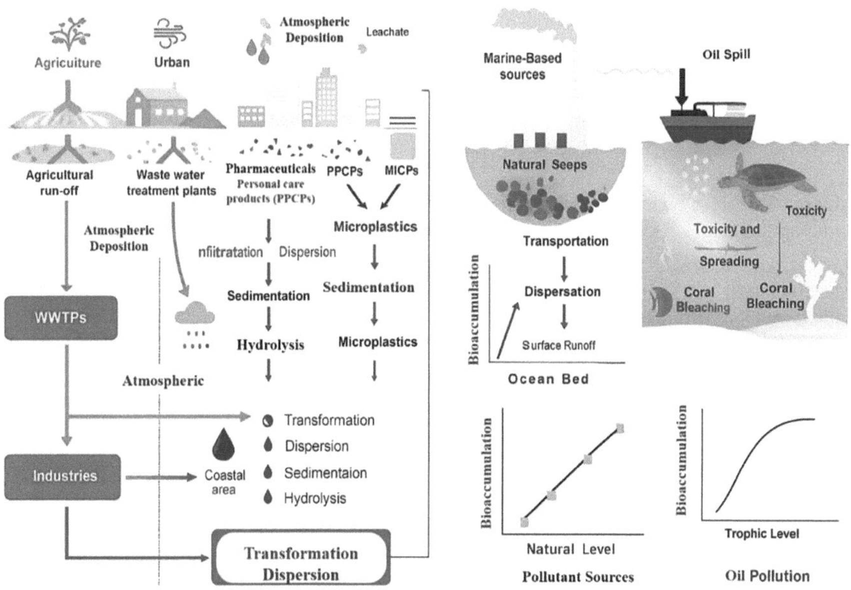

Fig. 10.1 Sources, pathways, and ecological impacts of pollutants in aquatic environments, including pharmaceuticals, microplastics, oil spills, and their ecological consequences including bioaccumulation, toxicity, and coral bleaching (Shankhwar et al. 2025a, b)

Figure 10.1 illustrates marine pollution from human and natural sources and its effects, including toxin accumulation, coral bleaching, and ecosystem disruption.

Tackling marine pollution, especially micropollutants, is tough because these contaminants come in so many forms, each behaving differently in the environment. The management of these substances requires scientists to understand their complete life cycle from origin to movement patterns and long-term environmental impact. The solution to this problem emerges through advanced scientific methods. By combining microbial processes with nuclear methods, scientists can develop new ways to track ocean pollution and find ways to reduce it.

10.1.2 Defining Marine Micropollutants: Scope and Classification

Marine micropollutants are compounds present in the water in minuscule concentrations, frequently below 1 microgram per liter (<1 μg/L), yet they can significantly impact ecosystems and human health (Warner et al. 2019). They often categorize into several primary groups:

Table 10.1 Classification and key characteristics of marine micropollutants

Category	Examples	Sources	Persistence	Potential impacts
Organic micropollutants	Pharmaceuticals, pesticides	Industrial discharge, agricultural runoff	High (years)	Endocrine disruption, reproductive harm
Inorganic micropollutants	Heavy metals (mercury, cadmium)	Mining, atmospheric deposition	Very high (decades)	Neurotoxicity, bioaccumulation
Plastic-based micropollutants	Microplastics, nanoplastics	Degraded consumer plastics, tire wear	High (centuries)	Physical harm, toxin transport *via* food web
Radionuclides	Cesium-137, Strontium-90	Nuclear accidents, natural seepage	Variable	Radiation poisoning, genetic mutations

Inorganic micropollutants: These pollutants are dominated by heavy metals such as lead (Pb), mercury (Hg), and cadmium (Cd), together with certain radionuclides, all of which are toxic even at trace concentrations (Abbasi et al. 2022).

Organic micropollutants: Cover a wide mix of chemicals, including pharmaceuticals, personal care products, pesticides, and synthetic industrial compounds (Ahmad 2022).

Plastic-based micropollutants: This includes micro and nanoplastics, which come from the breakdown of larger plastic waste (Sinha 2024; Shankhwar et al. 2025c), or everyday products like cosmetics and clothing.

Radionuclides: Radionuclides such as Cs-137 and Sr-90 enter marine ecosystems through nuclear accidents and natural seepage. Their longevity in the environment varies, but they can lead to radiation exposure and genetic mutations in marine organisms.

Table 10.1 outlines the different types of micropollutants, where they come from, how long they last, and the potential harm they can cause.

10.2 Marine Microbiomes and Their Role in Pollution Control

Micropollutants persist in the ocean even in small concentrations, as they break down slowly. Certain pharmaceuticals remain for months or years, continuously impacting marine organisms. PFAS and similar chemicals are designed for durability, making them difficult to eliminate from the environment. Out of the need to address micropollutants in the ocean, some focus on determining their impact on the microbiome (Shankhwar et al. 2025d). Microbial pollutants like hydrocarbons, tropospheric ozone pollutants, anthropogenic compounds, pharmaceutical wastes, heavy metals and MP modify microbial life in the oceans and hence their influence on the ecosystem is different and complicated (Shankhwar et al. 2025e). From a microbiological perspective, the ocean meiofauna like viruses, archaea, bacteria and other microorganisms along with surface eukaryotic phytoplankton (diatoms, dinoflagellates and flagellates) make up the marine microbiome. They are the constituents that ensure the stability of the ocean ecosystem and jointly maintain the health of the oceans by participating in the nutrient cycling, the degradation of pollutants, and other services provided to the oceans.

Figure 10.2 supports the importance of marine micropollutants sources like MP, pesticides, heavy metals, medicines, and even products of human body care industry. These pollutants are dangerous not only for marine life, but for the whole globe.

A better understanding of how micropollutants affect marine microbiomes is essential for improving pollution management (Tan et al. 2023). Micropollutants can disrupt these systems by altering microbial diversity, interfering with essential processes, and causing toxic effects (Shankhwar et al. 2025d). Marine microbiomes are diverse and active communities that help keep ocean ecosystems healthy (Pan and Wang 2023). Microbial biocatalysis involves the use of microorganisms or their enzymes to transform pollutants into safer, less toxic compounds (Mitra et al. 2022). Nuclear techniques such as radiolabeling are used to trace the movement and breakdown of pollutants, enabling the degradation process to be observed and analysed in

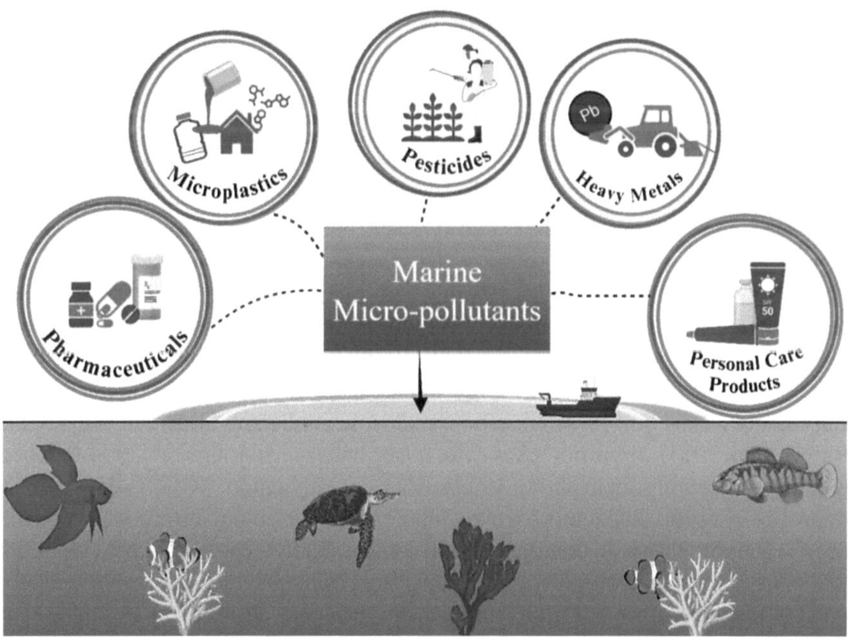

Fig. 10.2 Major contributors to marine micro-pollution: MP, pesticides, heavy metals, pharmaceuticals, and personal care products (Shankhwar et al. 2025a; Verma et al. 2025)

Table 10.2 Major microbial groups in marine microbiomes and their roles in pollution management

Microbial group	Main functions	Typical habitats	Roles in pollution management
Bacteria *Alcanivorax*, *Prochlorococcus* (Yakimov et al. 1998)	Decomposition, nutrient cycling, primary production	Surface waters, sediments, hydrothermal vents	Hydrocarbon degradation, heavy metal bioremediation
Archaea *Methanococcus*, *Nitrosopumilus* (Jarrell et al. 2021)	Methane production, ammonia oxidation	Deep-sea, hypersaline environments, sediments	Methane mitigation, metal ion transformations
Viruses Marine bacteriophages	Viral lysis of microbes, gene transfer	Open ocean, coastal marine environments	Modulation of microbial populations for pollutant degradation
Diatoms, dinoflagellates	Primary production, symbiotic relationships	Phytoplankton, zooplankton, coral reefs	Biodegradation of organic pollutants, nutrient cycling

real time. Table 10.2 summarizes the main microbial groups found in the ocean, their roles in the ecosystem, and how they could help in managing pollution.

10.3 The Diversity and Ubiquity of Marine Microbes in Pollution Management

Marine microbes are essential for breaking down pollutants because they can adapt to various conditions and have different ways of processing substances. Some microbes are tough enough to survive in extreme conditions and can break down stubborn pollutants; hydrocarbons and synthetic chemicals. *Alcanivorax borkumensis* was first identified by Yakimov and colleagues in 1998 from sediment in the North Sea (Yakimov et al. 1998). *Alcanivorax borkumensis* is particularly effective at degrading oil (Ahmad 2022). Its ability to break down hydrocarbons makes it a critical agent in bioremediation of oil-contaminated environments (Li et al. 2001; Lee et al. 2019). With tools like isotope tracing and radiolabeled tracking, scientists can closely follow how these microbes break down pollutants (Rostami et al. 2024; Rafeeq et al. 2023). Some microbes can even be enhanced through radiation-induced mutations, boosting their ability to clean up contamination. What makes marine microbes so useful is that they're everywhere; from the sunlit surface waters to the deep-sea floor. Many of them, especially gram-negative bacteria, have cell walls that naturally interact with oily, water-repelling pollutants, making them especially good at breaking down harmful substances in the ocean (Rostami et al. 2024; Pant et al. 2021). Scientists are learning more about how different marine bacteria are and, more crucially, how their genes and enzymes help break down contaminants thanks to high-throughput sequencing and metagenomics (Pant et al. 2021). When integrated with nuclear technologies, this understanding enables more targeted and effective approaches to mitigating marine pollution. By tapping into the natural abilities of microbes and enhancing their potential with advanced techniques, we can create sustainable ways to clean up marine environments, protect ecosystems, and minimize long-term environmental damage.

10.4 Sources and Pathways of Micropollutants

10.4.1 Runoff from Cities and Farms

Rain washes over streets, farmland, and industrial sites, carrying fertilizers, pesticides, pharmaceuticals, and even tiny plastic particles from worn-out tires into rivers and, eventually, the ocean (Matavos-Aramyan 2024; Dube and Okuthe 2023).

10.4.2 Pollution from the Air

Some pollutants travel long distances before settling in the ocean. Heavy metals and persistent organic chemicals get carried by the wind and fall into the water with rain (Hermabessiere et al. 2017; Mitra et al. 2022).

10.4.3 Direct Dumping

Waste from factories, untreated sewage, and oil spills go straight into the ocean, polluting both coastal waters and the open sea (Warner et al. 2019; Rostami et al. 2024).

Once in the ocean, these pollutants don't just disappear. Some sink into the sediment, building up over time, while others stay dissolved, spreading across vast areas. Whether they settle or drift, they can linger for years, affecting marine life and ecosystems.

10.5 Environmental and Ecological Impact

Endocrine Disruption: A variety of substances, industrial products and medications, can mimic or interfere with the hormones of marine creatures, causing endocrine disruption (Peivasteh-roudsari et al. 2023). Fish, mollusks, and crustaceans may experience developmental and reproductive problems as a result.

Biomagnification and Bioaccumulation: Pollutants are accumulated in marine animals and become more concentrated as they move through the food chain. As a result, high levels of these harmful substances can be found in top predators, such as marine mammals and humans. The risk of exposure is increased by consuming contaminated seafood, potentially leading to health issues such as nerve damage, reproductive disorders, and weakened immunity (Desiante et al. 2021).

Impact on Microbial Communities: Microbial enzyme activity can be interfered with by heavy metals, leading to disruptions in important biogeochemical cycles. Since these pollutants can persist in the environment for long periods and sometimes form toxic byproducts, careful monitoring and management are essential (Qian et al. 2020). Advanced analytical tools, including nuclear-based technologies to understand how these substances move and affect marine life (Mitra et al. 2022).

10.6 Nuclear-Based Approaches for Monitoring Marine Pollution

Nuclear instrumentation including nuclear spectroscopy, are applied to detect and quantify contaminants, radiometric analysis is used to measure and study pollutants, and isotopic tracing is employed to track contaminant sources, allowing accurate detection even at extremely low concentrations in the environment (Willis et al. 2022; Betti et al. 2011). Investigators use these methodologies to track pollutants accurately across different environments and over time. The movement and transformation of micropollutants are studied by applying isotopic tracers and radiolabeled substances, which provide detailed insights into their pathways and behavior in ecosystems. Observation techniques allow them to analyze how pollution spreads and to evaluate its broader environmental impacts (Betti et al. 2011; Anthony et al. 2023). Examining how pollutants interact with environmental elements like temperature, dissolved oxygen and salinity is especially insightful when these factors show significant fluctuations. Isotope Ratio Mass Spectrometry (IRMS) serves as an effective method for identifying the origins of contaminants, enabling analysts to pinpoint their sources with great precision and improving the comprehension of pollutant behavior in aquatic bodies. The technique reveals whether pollutants stem from natural or human-made sources which helps experts track their movements and changes (Kaklamanos et al. 2020). Radiometric techniques such as γ and alpha spectrometry, are used to detect and measure radionuclides and other contaminants in marine samples. Radiocarbon dating and related radiometric methods further allow researchers to determine the age of organic compounds and micropollutants. It provides valuable insights into their sources and pathways within the aquatic environment. It helps ecologists understand how pollution patterns change over time (Garcia-Tenorio et al. 2020). The persistence of toxins in deep-sea sediments over decades, as shown by radiocarbon dating, underscores the importance of ongoing pollution monitoring and long-term management approaches (Heim and Schwarzbauer 2013). In a study, the application of isotopic tracing shows that particular microbial groups possess the ability to break down contaminants and radiolabeled polycyclic aromatic hydrocarbons (PAHs) (Alaidaroos 2023). By tracking contaminant dynamics, researchers can determine the most effective microbial strains and environmental conditions for bioremediation. The application of γ spectrometry in this context enhances monitoring efforts by detecting pollutants at extremely low concentrations, enabling early identification of harmful substances before they exert significant ecological impacts (Anthony et al. 2023) (Kaklamanos et al. 2020). The approach aids in conserving biodiversity, conducting environmental surveillance and maintaining human health within marine environments.

10.7 Microbial Biocatalysis and Nuclear Techniques

Integration of nuclear techniques and microbial remediation aids in monitoring the degradation of pollutants and improving and tracking remediation approaches of bioremediation over time. By tracking the availability of key nutrients such as phosphorus and nitrogen, radiotracers reveal locations where microbial activity is constrained, which can slow the degradation of contaminants. Radiotracers help measure key water conditions like pH, oxygen, temperature, and salinity, which affect microbial activity and bioremediation success (Hassan et al. 2022; Rodrigues and de Carvalho 2022). Radiolabeling is a critical technique for monitoring biodegradation processes, enabling researchers to track the fate and behavior of specific substances in environmental systems. By attaching radioactive isotopes to target compounds, scientists can observe their movement and transformation with high precision, using specialized instruments to detect emitted radiation. This technique shows how marine microorganisms break down micropollutants and helps identify the best strains and conditions for effective bioremediation (Hassan et al. 2022). Using isotopic analysis in conjunction with microbial activity studies is a great way to gather information over the long-term regarding pollution. If the microbial research shows that there is a reduction in the efficiency of biodegradation because of environmental stress, then it is possible to react by nutrient addition or habitat improvement. If isotopic analysis indicates a persistent pollution source, regulatory measures can be adjusted to address it more effectively (Pan and Wang 2023; Hassan et al. 2022). Nuclear-based techniques for evaluating bioavailability, tracking pollutants, and analyzing microbial interactions in aquatic settings are shown in Fig. 10.3.

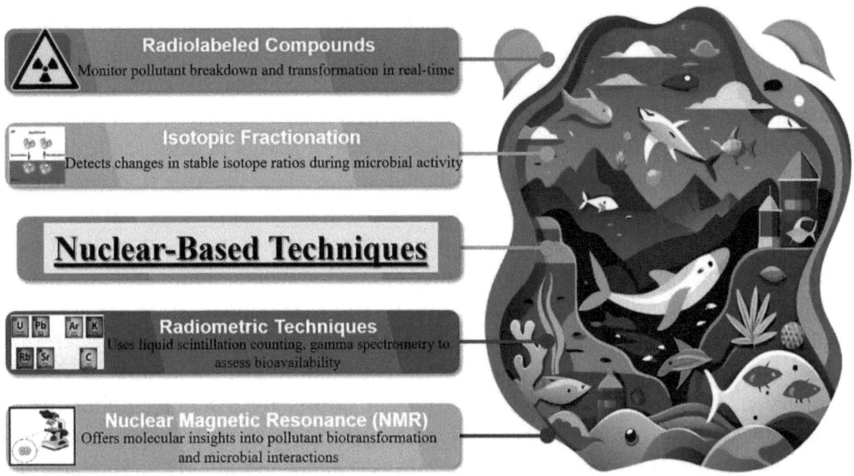

Fig. 10.3 Nuclear-based approaches: NMR, radiolabeled compounds, isotopic fractionation, and radiometric approaches for analyzing environmental pollutants

10.8 Tracking Microbial Degradation Using Radiolabeled Compounds

Pollutants with isotopes ^{35}S, tritium, or ^{14}C, can be radiolabelled and their microbial breakdown processes can be precisely tracked. Changes in radioactivity linked to various environmental compartments; water, sediment, or biota, can be used to track the breakdown and transformation of pollutants that have been radiolabeled (Garcia-Tenorio et al. 2020; Touranlou et al. 2024). The efficiency of breakdown and microbial activity can be directly measured using this method.

10.8.1 Radiolabeled Substrates for Biodegradation Studies

Pesticides, hydrocarbons, and pharmaceuticals can be radiolabeled to monitor their breakdown by microbial communities. ^{14}C-labeled naphthalene being used to study how Pseudomonas and Mycobacterium degrade polycyclic aromatic hydrocarbons (PAHs) (Zeng et al. 2021). The amount of mineralization can be measured, which shows that the molecule has completely broken down into non-toxic end products (CO_2 and H_2O), by monitoring the emission of $^{14}CO_2$ during biodegradation.

10.8.2 Utilizing Radiolabeled Pollutants for Biotransformation Research

Radiolabeling allows for the detection of intermediate metabolites formed during microbial biotransformation. By tracking the breakdown of ^{14}C-labeled chlorinated compounds, dechlorination byproducts such as dichloroethylene or vinyl chloride can be identified (Desiante et al. 2021). This knowledge helps refine bioremediation strategies to prevent the buildup of toxic intermediates during the treatment process.

10.8.3 Isotopic Fractionation as a Biodegradation Diagnostic Method

Isotopic fractionation occurs because isotopes of the same element have slightly different masses, causing them to participate in chemical and biological reactions at different rates. In microbial degradation of organic contaminants, lighter isotopes (^{12}C) are often processed more quickly than heavier isotopes (^{13}C), leading to an enrichment of the heavier isotope in the remaining substrate. During nitrogen cycling, processes like denitrification and nitrogen assimilation can alter $^{15}N/^{14}N$ ratios, providing insight into nutrient sources and transformations. The development and mechanisms of microbial biodegradation in marine environments can be assessed using this phenomenon as a diagnostic technique.

10.8.4 Stable Isotope Analysis for Tracking Biodegradation

Microorganisms often prefer lighter isotopes, such as ^{12}C over ^{13}C, during the breakdown of pollutants. This process leads to a higher concentration of heavier isotopes in the residual contaminant. These isotopic variations using IRMS, biodegradation can be evaluated, even when overall pollutant concentrations remain constant (Betti et al. 2011; Kaklamanos et al. 2020). Elevated levels of ^{13}C in residual oil fractions serve as a sensitive indicator of hydrocarbon degradation following an oil spill. Variations in the $^{15}N/^{14}N$ ratio in nitrogen-containing pollutants provide evidence of bacterial denitrification activity, offering insights into microbial processing of contaminants in affected marine environments.

10.8.5 Monitoring Microbial Activity Using Radiometric Techniques

Liquid scintillation counting and γ spectrometry are employed to quantify the extent of pollutant degradation and to evaluate the bioavailability of contaminants for microbial uptake. These techniques enable precise monitoring of the absorption and mineralization of radiolabeled pollutants by microorganisms, providing critical insights into the efficiency and dynamics of bioremediation processes. The processes are investigated to gain a better understanding of microbial processes and to determine the efficiency of bioremediation techniques (DeBlanc et al. 1972).

10.8.6 Compound-Specific Isotope Analysis (CSIA)

CSIA is a method that allows us to monitor the natural degradation of pollutants through the analysis of the isotopic composition of the contaminants and their reaction products. This method allows to distinguish between the microbial degradation of the contaminants and other factors such as light. Trichloroethylene (TCE) is a common pollutant containing chlorine. Changes that signify microbial dechlorination can be found by examining the carbon and chlorine isotopes in a chemical, which aids in identifying the particular bacterium causing the breakdown (Zhang 2017).

10.8.7 Utilizing Radiotracers to Evaluate Bioavailability

Radiotracers help track how pollutants interact with environmental elements like organic matter and sediments. Understanding how pollutants shift between solid and liquid phases is key to improving desorption methods (Touranlou et al. 2024; DeBlanc et al. 1972). This helps make contaminants more accessible to microorganisms for effective breakdown. Radiotracers are used to track how ^{14}C-labeled hydrophobic organic compounds, like PAHs, attach to marine sediments.

10.8.8 Utilizing Gamma (γ) Spectrometry to Track Radionuclide Bioremediation

During microbial bioremediation of radioactive contaminants, γ spectrometry is used to detect changes in radionuclide levels, allowing the progress of the cleanup process to be tracked (Islami Rad et al. 2023). Soluble radionuclides can be converted into insoluble forms, like the reduction of UO_2^{2+} to UO_2, allowing them to settle and be removed from water. This process is carried out by microbes such as *Shewanella* species, which thrive in marine environments, and *Geobacter* species, which play a key role in metal reduction (Islami Rad et al. 2023).

10.9 Microbial Metabolism Research Using Nuclear Magnetic Resonance (NMR) Spectroscopy

NMR helps detect alcohols, carboxylic acids, and other byproducts formed when microorganisms break down complex hydrocarbons. This shows the various biochemical pathways utilized by different bacterial species in the process (Peng et al. 2024). Using NMR spectroscopy, we can observe how pollutants change their structure as they're transformed by living organisms. This technique helps to understand how microorganisms metabolize pollutants, showing the metabolites of the process as well as the end products (Peng et al. 2024; Moco 2022).

10.9.1 Analyzing the Interactions Between Pollutants and Microbe

NMR spectroscopy serves as a fundamental tool for investigating how microbial cell membranes interact with pollutants. This technique provides detailed insights into the extent of pollutant-induced damage and reveals the mechanisms by which bacteria internalize and degrade contaminants through their membranes. When applied to samples contaminated with heavy metals, NMR spectroscopy allows researchers to examine alterations in microbial metabolic activity and stress responses. Insights gained from these analyses inform strategies to enhance microbial resilience and optimize their capacity for pollutant transformation, ultimately improving the effectiveness of bioremediation efforts (Moco 2022; Savorani et al. 2013). In environmental cleanup, NMR spectroscopy is helpful in understanding pollutant transformation and microbial metabolism by providing key structural insights.

10.9.2 Bioaugmentation and Biostimulation

Bioremediation is a naturally occurring process in which microorganisms degrade harmful pollutants in soil, water, and air. The efficiency of this process depends on environmental conditions, including nutrient availability, temperature, oxygen concentration, and pH. Under optimal conditions, bacteria, fungi, and other microbes metabolize toxins, transforming them into harmless byproducts such as CO_2 and water, thereby reducing the environmental impact of contamination. Biostimulation is enhancing environmental conditions to make available nutrients to existing microorganisms to increase their metabolic rates to use for degrading pollutants. In contrast, bioaugmentation involves the addition of specific microbial strains that are responsible for the degradation of specific contaminants, and thus is a more controlled and better method of treatment. Although both techniques are usually applied in combination to enhance the remediation, bioaugmentation is usually faster in terms of degradation rates and more effective in treatment. This is because it can target pollutants that cannot be easily removed by biostimulation. Bioaugmentation is useful in the treatment of varied pollutants such as petroleum hydrocarbons, heavy metals and chlorinated solvents and therefore has a great potential in multi contaminated sites. It improves the metabolic function of the microbes by provision of nutrients or electron acceptor(s) whether or not bioaugmentation is used. Bioaugmentation is a more specific method of pollutant removal than biostimulation which acts to enhance growth of microbes. The use of both techniques in combination results in improved and sustainable cleanup plans that increase the rate of pollution removal, thus protecting human health and the environment (Essabri et al. 2019). Nuclear techniques are used to assess bioremediation by tracking changes in the isotopic composition, the rate at which microbes incorporate pollutants and the total levels of pollutants (Radhakrishnan et al. 2023).

10.9.3 Tracking the Long-Term Results of Remediation

Microbial analysis, combined with nuclear techniques, is used to detect recontamination, identify transformation products, and assess residual contamination (Peng et al. 2024). Isotopic studies and radiometric surveys are carried out to determine concentrations of pollutants and their changes in composition to determine whether there is a need for further treatment following bioremediation. Combining microbial cleanup methods with nuclear techniques gives scientists a better understanding of how pollutants behave. This approach helps create better strategies for tackling marine pollution (Pandey et al. 2009).

Table 10.3 demonstrates the synergy between the two techniques when used together for monitoring and remediation purposes.

Table 10.3 Nuclear methods with microbial biocatalysis to manage marine pollution

Application area	Microbial role	Nuclear technique	Outcome
Pollutant degradation	Microbial breakdown of PAHs, pesticides	Radiolabeling (e.g., ^{14}C)	Tracking degradation pathways, quantifying breakdown efficiency
Bioavailability assessment	Uptake of pollutants by microbial communities	γ spectrometry, IRMS	Identifying pollutant accessibility for microbes
Optimizing bioremediation	Enhancing conditions for microbial activity	Isotopic analysis, radiometric tools	Fine-tuning nutrient levels, salinity, and pH for maximum degradation
Long-term monitoring	Sustaining microbial degradation processes	Radiometric surveys, isotopic tracking	Assessing remediation success and pollutant re-distribution trends

10.10 Conclusions

Ocean life and human health are seriously at risk from marine pollution, which is brought on by persistent pollutants. In constantly changing marine environments, these pollutants are difficult to control using traditional cleanup methods, making pollution management more challenging. Combining microbial bioremediation with nuclear techniques provides a powerful approach to tracking, understanding, and reducing contamination. Natural microbes break down pollutants, however nuclear methods help to detect changes at molecular level, and thus provide very detailed monitoring and analysis of the cleanup operation. Using nuclear techniques to improve microbial degradation speeds up bioremediation, leading to faster pollutant removal and fewer toxic byproducts. Accurate tracking of tiny pollutants and their breakdown products in the ocean is enabled by nuclear techniques such as radiolabeling and spectrometry. This improves the assessment of contamination and provides real time monitoring of marine cleanup. They provide flexible solutions for both rapid pollution response and continuous environmental monitoring, making them essential for effective contamination management. Being able to watch how pollutants act helps to act early and sustainable way of pollution management is better. This is why it is important to combine nuclear technologies with microbial degradation in order to protect marine ecosystems and maintain their natural balance. Enhancing these methods will be crucial to preserving marine habitats as environmental problems continue to escalate. Effective marine and conservation will also depend heavily on the adoption of supportive policies and cross-sector collaboration.

Acknowledgments The author expresses sincere gratitude to the Department of Biotechnology (DBT), Ministry of Science & Technology, Government of India, for their generous funding under Project No. DBT-RA/2022/July/N/2240 through the DBT-RA Program. Special thanks go to Prof. Aynur Unal and Dr. Mohit Kumar for their valuable insights and contributions to improving this study. Finally, a heartfelt thanks to my daughter, Vanshika, whose love and constant curiosity are my greatest inspiration and source of strength.

References

Abbasi NA, Shahid SU, Majid M, Tahir A (2022) Ecotoxicological risk assessment of environmental micropollutants. In: Environmental micropollutants. Elsevier, pp 331–337. https://doi.org/10.1016/B978-0-323-90555-8.00004-0

Ahmad I (2022) Microalgae–Bacteria Consortia: a review on the degradation of polycyclic aromatic hydrocarbons (PAHs). Arab J Sci Eng 47(1):19–43. https://doi.org/10.1007/s13369-021-06236-9

Alaidaroos BA (2023) Advancing eco-sustainable bioremediation for hydrocarbon contaminants: challenges and solutions. Processes 11(10):3036. https://doi.org/10.3390/pr11103036

Anthony D et al (2023) Trends in marine pollution mitigation technologies: scientometric analysis of published literature (1990–2022). Reg Stud Mar Sci 66:103156. https://doi.org/10.1016/j.rsma.2023.103156

Baek S-S, Yun D, Pyo J, Kang D, Cho KH, Jeon J (2022) Analysis of micropollutants in a marine outfall using network analysis and decision tree. Sci Total Environ 806:150938. https://doi.org/10.1016/j.scitotenv.2021.150938

Betti M, Boisson F, Eriksson M, Tolosa I, Vasileva E (2011) Isotope analysis for marine environmental studies. Int J Mass Spectrom 307(1–3):192–199. https://doi.org/10.1016/j.ijms.2011.03.008

Bilal M, Lam SS, Iqbal HMN (2022) Biocatalytic remediation of pharmaceutically active micropollutants for environmental sustainability. Environ Pollut 293:118582. https://doi.org/10.1016/j.envpol.2021.118582

DeBlanc HJ, Charache P, Wagner HN (1972) Automatic radiometric measurement of antibiotic effect on bacterial growth. Antimicrob Agents Chemother 2(5):360–366. https://doi.org/10.1128/AAC.2.5.360

Desiante WL, Minas NS, Fenner K (2021) Micropollutant biotransformation and bioaccumulation in natural stream biofilms. Water Res 193:116846. https://doi.org/10.1016/j.watres.2021.116846

Dube E, Okuthe GE (2023) Plastics and micro/nano-plastics (MNPs) in the environment: occurrence, impact, and toxicity. Int J Environ Res Public Health 20(17):6667. https://doi.org/10.3390/ijerph20176667

Essabri AMA, Aydinlik NP, Williams NE (2019) Bioaugmentation and biostimulation of total petroleum hydrocarbon degradation in a petroleum-contaminated soil with fungi isolated from olive oil effluent. Water Air Soil Pollut 230(3):76. https://doi.org/10.1007/s11270-019-4127-8

Garcia-Tenorio R et al (2020) From radiometry to chronology of a marine sediment core: a ^{210}Pb dating interlaboratory comparison exercise organised by the IAEA. Mar Pollut Bull 159:111490. https://doi.org/10.1016/j.marpolbul.2020.111490

Gaw S, Thomas KV, Hutchinson TH (2014) Sources, impacts and trends of pharmaceuticals in the marine and coastal environment. Philos Trans R Soc Lond B Biol Sci 369(1656):20130572. https://doi.org/10.1098/rstb.2013.0572

Hassan S et al (2022) Marine bacteria and omic approaches: a novel and potential repository for bioremediation assessment. J Appl Microbiol 133(4):2299–2313. https://doi.org/10.1111/jam.15711

Heim S, Schwarzbauer J (2013) Pollution history revealed by sedimentary records: a review. Environ Chem Lett 11(3):255–270. https://doi.org/10.1007/s10311-013-0409-3
Hermabessiere L et al (2017) Occurrence and effects of plastic additives on marine environments and organisms: a review. Chemosphere 182:781–793. https://doi.org/10.1016/j.chemosphere.2017.05.096
Islami Rad SZ, Mansuri R, Rezaei GH (2023) Determination of natural radioactivity using gamma spectrometry in used soils as construction materials in Qom province, Iran. J Radiat Res Appl Sci 16(4):100753. https://doi.org/10.1016/j.jrras.2023.100753
Jarrell KF, Albers S-V, de S. Machado JN (2021) A comprehensive history of motility and Archaellation in Archaea. FEMS Microbes 2. https://doi.org/10.1093/femsmc/xtab002
Kaklamanos G, Aprea E, Theodoridis G (2020) Mass spectrometry: principles and instrumentation. In: Chemical analysis of food. Elsevier, pp 525–552. https://doi.org/10.1016/B978-0-12-813266-1.00011-5
Lee Y, Lee Y, Jeon CO (2019) Biodegradation of naphthalene, BTEX, and aliphatic hydrocarbons by Paraburkholderia aromaticivorans BN5 isolated from petroleum-contaminated soil. Sci Rep 9(1):860. https://doi.org/10.1038/s41598-018-36165-x
Li Q-S, Ogawa J, Schmid RD, Shimizu S (2001) Engineering cytochrome P450 BM-3 for oxidation of polycyclic aromatic hydrocarbons. Appl Environ Microbiol 67(12):5735–5739. https://doi.org/10.1128/AEM.67.12.5735-5739.2001
Matavos-Aramyan S (2024) Addressing the microplastic crisis: a multifaceted approach to removal and regulation. Environ Adv 17:100579. https://doi.org/10.1016/j.envadv.2024.100579
Mitra S et al (2022) Impact of heavy metals on the environment and human health: novel therapeutic insights to counter the toxicity. J King Saud Univ Sci 34(3):101865. https://doi.org/10.1016/j.jksus.2022.101865
Moco S (2022) Studying metabolism by NMR-based metabolomics. Front Mol Biosci 9. https://doi.org/10.3389/fmolb.2022.882487
Pan T, Wang Z (2023) Microbial biocatalysis. Catalysts 13(3):629. https://doi.org/10.3390/catal13030629
Pandey J, Chauhan A, Jain RK (2009) Integrative approaches for assessing the ecological sustainability of in situ bioremediation. FEMS Microbiol Rev 33(2):324–375. https://doi.org/10.1111/j.1574-6976.2008.00133.x
Pant G, Garlapati D, Agrawal U, Prasuna RG, Mathimani T, Pugazhendhi A (2021) Biological approaches practised using genetically engineered microbes for a sustainable environment: a review. J Hazard Mater 405:124631. https://doi.org/10.1016/j.jhazmat.2020.124631
Peivasteh-roudsari L et al (2023) Origin, dietary exposure, and toxicity of endocrine-disrupting food chemical contaminants: a comprehensive review. Heliyon 9(7):e18140. https://doi.org/10.1016/j.heliyon.2023.e18140
Peng Y, Zhang Z, He L, Li C, Liu M (2024) NMR spectroscopy for metabolomics in the living system: recent progress and future challenges. Anal Bioanal Chem 416(9):2319–2334. https://doi.org/10.1007/s00216-024-05137-8
Qian X et al (2020) Biotechnological potential and applications of microbial consortia. Biotechnol Adv 40:107500. https://doi.org/10.1016/j.biotechadv.2019.107500
Radhakrishnan A et al (2023) Bioremediation of hydrocarbon pollutants: recent promising sustainable approaches, scope, and challenges. Sustainability 15(7):5847. https://doi.org/10.3390/su15075847
Rafeeq H et al (2023) Genetically engineered microorganisms for environmental remediation. Chemosphere 310:136751. https://doi.org/10.1016/j.chemosphere.2022.136751
Rodrigues CJC, de Carvalho CCCR (2022) Marine bioprospecting, biocatalysis and process development. Microorganisms 10(10):1965. https://doi.org/10.3390/microorganisms10101965
Rostami M, Ghorbani A, Shahbazi S (2024) Gamma radiation-induced enhancement of biocontrol agents for plant disease management. Curr Res Microb Sci 7:100308. https://doi.org/10.1016/j.crmicr.2024.100308

Savorani F, Rasmussen MA, Mikkelsen MS, Engelsen SB (2013) A primer to nutritional metabolomics by NMR spectroscopy and chemometrics. Food Res Int 54(1):1131–1145. https://doi.org/10.1016/j.foodres.2012.12.025

Shankhwar N, Verma AK, Shaik S, Pnadita K (2025a) Microbes as biocatalysts of marine micropollutants. In: Marine microbiomes: exploring ecology, biotechnology, and environmental. CRC Press, Taylor and Francis

Shankhwar N, Verma AK, Rao KS, Vivekanand (2025b) Symbiotic relationships in marine microbial communities. In: Marine microbiomes: exploring ecology, biotechnology, and environmental. CRC Press, Taylor and Francis

Shankhwar N, Singh P, Thomas J (2025c) Chapter 6: Nanocomposites as smart nanomaterials for photodegradation of micro and nano-plastics. In: Kumar S (ed) Emerging materials for photodegradation and environmental remediation of micro- and nano-plastics, 1st edn. Springer Nature

Shankhwar VN, Verma AK, Rao KS (2025d) Symbiotic relationships in marine microbial communities. In: Marine microbiomes: exploring ecology, biotechnology, and environmental impact. CRC Press, Taylor and Francis

Shankhwar KPN, Verma AK, Shaik S (2025e) Microbes as biocatalysts of marine micropollutants. CRC Press, Taylor and Francis

Sinha S (2024) Plastic debris, microplastics, and marine pollution. In: Plastic pollution. Springer Nature Singapore, Singapore, pp 79–98. https://doi.org/10.1007/978-981-97-5528-8_5

Spindola Vilela CL, Damasceno TL, Thomas T, Peixoto RS (2022) Global qualitative and quantitative distribution of micropollutants in the deep sea. Environ Pollut 307:119414. https://doi.org/10.1016/j.envpol.2022.119414

Tan WA, Celina G, Pranawijaya S (2023) Bioremediation in the marine environment: challenges and prospective methods for enhancement. E3S Web Conf 374:00038. https://doi.org/10.1051/e3sconf/202337400038

Touranlou FA, Moghimani M, Marhamati M, Rezaei M (2024) Detection and measurement of radioactive substances in water and food: a narrative review. Ital J Food Saf. https://doi.org/10.4081/ijfs.2024.11651

Verma AK, Pawan A, Kathpalia KV, Singh T, Shankhwar N (2025) Marine microbiomes and biogeochemical cycles. In: Marine microbiomes: exploring ecology, biotechnology, and environmental impact. CRC Press, Taylor and Francis

Warner W, Licha T, Nödler K (2019) Qualitative and quantitative use of micropollutants as source and process indicators. A review. Sci Total Environ 686:75–89. https://doi.org/10.1016/j.scitotenv.2019.05.385

Willis KA et al (2022) Cleaner seas: reducing marine pollution. Rev Fish Biol Fish 32(1):145–160. https://doi.org/10.1007/s11160-021-09674-8

Yakimov MM et al (1998) Alcanivorax borkumensis gen. nov., sp. nov., a new, hydrocarbon-degrading and surfactant-producing marine bacterium. Int J Syst Bacteriol 48(2):339–348. https://doi.org/10.1099/00207713-48-2-339

Zeng J, Li Y, Dai Y, Wu Y, Lin X (2021) Effects of polycyclic aromatic hydrocarbon structure on PAH mineralization and toxicity to soil microorganisms after oxidative bioremediation by laccase. Environ Pollut 287:117581. https://doi.org/10.1016/j.envpol.2021.117581

Zhang YG (2017) Compound-specific isotope analysis. Springer, pp 1–4. https://doi.org/10.1007/978-3-319-39193-9_92-1

Chapter 11
Nuclear Energy for Desalination in Seawater: A Solution for Arid Regions?

Khadija El-Moustaqim, Chaimae Lamiri, Dalal Boudoudou, Karima Azoulay, Imane Bencheikh, Aziza Lamchaimech, and Jamal Mabrouki

Abstract The urgent problem of potable water scarcity, especially in arid and semi-arid regions of the world, has a possible solution in nuclear desalination. This technology provides a sustainable substitute for existing water supply methods by using the thermal energy produced by nuclear reactors, which is typically regarded as a byproduct, to desalinate saltwater. To address this growing need and guarantee that communities have access to vital water supplies for agriculture, drinking, and sanitation, nuclear desalination is a feasible alternative. Moreover, climate change intensifies water scarcity by altering conventional precipitation patterns, resulting in droughts and floods across different areas. This erratic unpredictability makes nuclear desalination a dependable supply of freshwater, ensuring water availability in areas susceptible to climate-induced shortages. In this article, we have conducted a study to evaluate the current status of desalination with nuclear energy by assessing the innovative technologies for seawater desalination, which could be coupled to the main types of existing nuclear energy. It also provides the recent developments in nuclear desalination and its future role are discussed.

Keywords Nuclear desalination systems · Seawater · Water · Arid region · Nuclear reactors

---

K. El-Moustaqim (✉)
Improvement and Valuation of Plant Resources, Faculty of Sciences, Ibn Tofaïl University—KENITRA-University Campus, Kenitra, Morocco

C. Lamiri · D. Boudoudou · K. Azoulay · I. Bencheikh · J. Mabrouki
Laboratory of Spectroscopy, Molecular Modelling, Materials, Nanomaterials, Water and Environment, CERN2D, Faculty of Science, Mohammed V University in Rabat, Rabat, Morocco

A. Lamchaimech
LG2E Laboratory, Department of Geology, Faculty of Science, Mohammed V University in Rabat, Rabat, Morocco

N. Mejjad (ed.), *Nuclear-Based Approaches application for marine pollution assessment and monitoring*, Environmental Science and Engineering,
https://doi.org/10.1007/978-3-032-12197-4_11

11.1 Introduction

The worldwide supply of drinkable water greatly exceeds the necessary and utilized quantities; nonetheless, resource allocation is still inequitable (El-Moustaqim et al. 2023a, b). There are areas where water is limited and the population is vulnerable to insufficient resources (El-Moustaqim et al. 2024a, b; Tallou et al. 2024). Seawater is a virtually inexhaustible resource (El-Moustaqim et al. 2025a, b). The generation of drinkable water through new strategic cooperation frameworks manifests as a significant aspect of the region, potentially aiding in the protection and maintenance of water resources (Assia and Mohamed 2024; Yahyaoui and Chibani 2024). This participatory framework may perhaps guarantee long-term security for numerous nations. A cost-effective way to produce potable water is implementing nuclear-powered desalination technology (Ali et al. 2024; Kamboj et al. 2020; Lønborg et al. 2024; Rich 2011). The development, provision, conservation, and preservation of adequate water supplies have emerged as a crucial issue. It is evident how crucial it is to explore the feasibility of, and backing for, saltwater desalination via nuclear energy. Furthermore, desalination powered by fossil fuels would be incompatible with sustainable development (Belhaj et al. n.d.; Benaafi et al. 2024). Growing worries about the expense and unpredictability of fossil fuels and related environmental issues have spurred a search for alternative energy sources, such as nuclear energy, for future desalination demands (Sara and Nacera 2015; Tahri et al. n.d.). Because of environmental worries about the rising concentration of greenhouse gases (GHGs), especially carbon dioxide, in the atmosphere, nuclear seawater desalination is becoming more attractive than traditional technologies (Wu et al. 2024). Desalination uses energy, much like any other industrial operation. However, the depletion of fossil fuels and their growing intensity of usage creates environmental issues, such as the possibility of a slow but significant shift in climate. Simultaneously, global energy consumption is consistently increasing, necessitating effective solutions (Buzzetti et al. 2024; Faibish et al. n.d.). Instead, by reducing the environmental impact and removing some of the barriers to using renewable energy sources, nuclear desalination can greatly aid in reaching sustainability goals. Three technologies are used in nuclear desalination: nuclear, desalination, and their connection system (Al-Othman et al. 2019; Misra and Kupitz 2004). Nuclear desalination is the process of turning saltwater into drinkable water using energy from a nuclear reactor. Nuclear reactors are not specifically chosen for desalination technology; any reactor can be connected to a desalination system. The process of creating fresh water with an on-site nuclear reactor is known as nuclear desalination (Wu et al. 2024; Al-Othman et al. 2019).

The amount of nuclear desalination-related research that has been published in the literature is growing. Therefore, this chapter's goal is to present a thorough summary of the most recent research on the several facets of nuclear desalination. Additionally, it aims to assess the present hybrid trends in desalination as well as the next research endeavors. In this context, the evaluation of safety issues and economic impact is also provided.

11.2 Nuclear Energy

With an 81% contribution, fossil fuels have dominated the energy landscape for the past century in both developed and industrialized nations. However, due to significant investments in sustainable and renewable energy sources, the rate of energy consumption has changed during the past 15 years (Ritchie et al. 2024). Particularly, nuclear power attracted a lot of attention. This is brought on by the need to expand fuel sources, boost the necessary energy supply, and reduce reliance on oil imports and non-renewable energy sources. These elements coincide with several environmental issues, including air pollution, climate change, and the greenhouse effect, that are brought on by the overuse of fossil fuels as main energy sources (Saleh et al. 2024; Wang et al. 2024). The four main factors that affect nuclear power generation are capital costs, operation and maintenance expenses, fuel prices, and back-end costs. These factors have to do with the decommissioning and disposal of end-of-life plants. Due to the low cost of fuel, assessments and sensitivity studies about the generation of energy have demonstrated that, in many markets, nuclear power is the most affordable option for supplying electricity (Wang et al. 2024; Ozcan et al. 2024).

In order to counteract unstable and high oil prices and lessen reliance on imported fossil fuels, nuclear energy is anticipated to be a significant alternative source. As a result, nuclear energy, a cheap, accessible, and carbon-free energy source, may prove to be a crucial substitute for fossil fuels (Ozcan et al. 2024). Nevertheless, while being a much sought-after alternative source, nuclear energy continues to hold a vague place in the literature today. However, proponents of nuclear energy assert that, in contrast to fossil fuels, nuclear power plants have cheap running costs but high installation expenses. On a worldwide scale, everyone agrees that using nuclear power for desalination is both feasible and profitable. An overview of desalination technology and the various nuclear reactors frequently utilized in desalination are illustrated in Fig. 11.1 (Bozkaya et al. 2025).

11.3 Nuclear Desalination in the MENA Countries

Nuclear reactor technology and its prospective applications in desalination and energy cogeneration are of interest to the majority of Middle Eastern nations. Countries in the MENA region, which is desert, are strongly motivated to employ nuclear energy to meet their electricity and water needs due to the region's rapid industrialization and population increase (Saleh et al. 2024). Countries in arid regions are hesitant to harness this energy source due to the high level of technological development required (Kamboj et al. 2020; Lønborg et al. 2024). But lately, nations like Kuwait, the United Arab Emirates, and the Kingdom of Saudi Arabia have rekindled their interest in the peaceful nuclear program. As a result, research into the possibilities of nuclear energy in desalination and electrical cogeneration has begun in these nations (Abdel Monem et al. 2024). According to international

Fig. 11.1 Application of Nuclear energy in the seawater

data, countries in the Middle East and North Africa (MENA) are more vulnerable to water scarcity than other regions of the world. According to a number of measures, 14 MENA countries rank among the top 20 water-scarce countries. Because of the aforementioned factors, it is therefore expected that the outlook for fresh water would worsen. For example, scientists predict that the Middle East's allowable freshwater per capita will be half of what it is now by 2025 (El-Ashmawy et al. 2024). Right now, the only substantial unconventional method for locating an affordable water supply is the desalination process. This is the only technically viable method on a broad scale. There is no denying that nuclear power is superior to all other energy sources. Because it is seen as a low-carbon or greenhouse gas emission-free power plant, it is therefore thought to be highly competitive. By supplying both low pressure and temperature steam to the industrial processes, heat from the power plant can be obtained at varying temperatures in the form of hot water, steam, or hot gas (Raymond et al. 2024).

11.4 Combining Power Generating and Water Desalination

Improvements in safety and technology advancements are driving the linking of nuclear reactors with desalination plants. Both desalination and nuclear (for energy) are unquestionably well-established technologies (see Fig. 11.2). Thorough study on both technologies over the last few decades has made this possible. Over 11.5% of the world's electricity is thought to come from nuclear technology (Abdel Monem et al. 2024; Raymond et al. 2024). Numerous nuclear reactor connections to small and medium-sized desalination plants have been documented in the literature. Advanced research opportunities are being created by the rapid advancements in the fields of nuclear reactors and desalination technologies. The direction of future potable water production will be influenced by coupling process optimization, which will also establish the role of sustainable energy in the creation of "new water" (Zhong et al. 2024).

The development of novel reactor concepts and fuel cycles with improved safety features, which are anticipated to increase public acceptance, the production of less expensive energy compared to other options, the need for prudent use of fossil energy sources, and the growing requirements to reduce the production of greenhouse gases (GHG), toxic gases, particulates, and acid rain all of which are associated with the combustion of fossil fuels are some of the many factors that support a potential resurgence of nuclear power production in the years to come (Wu et al. 2024; Wang et al. 2024; Bozkaya et al. 2025).

Nuclear energy can be a viable source of energy for seawater desalination, just as it may be for the production of electricity. The process of creating drinkable water from saltwater in an integrated facility using a nuclear reactor as the energy source (steam and/or electricity) is known as nuclear desalination (Fig. 11.3). Potable water production may be the only purpose of the facility, or it may be used to generate electricity and produce potable water, in which case only a part of the reactor's total

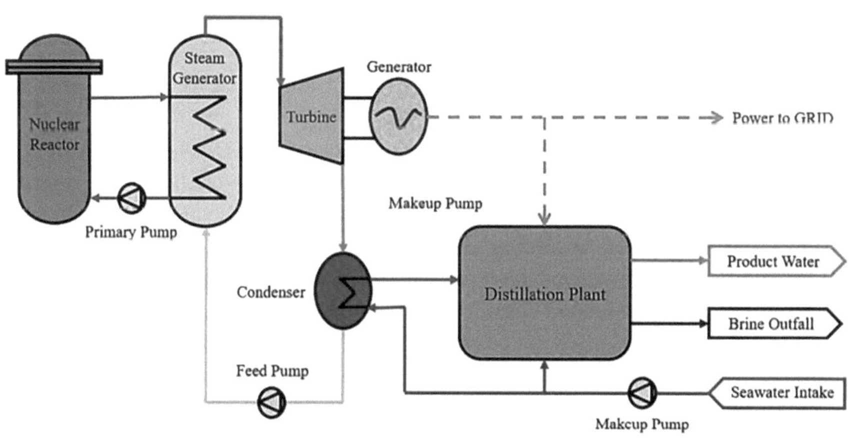

Fig. 11.2 Nuclear desalination process

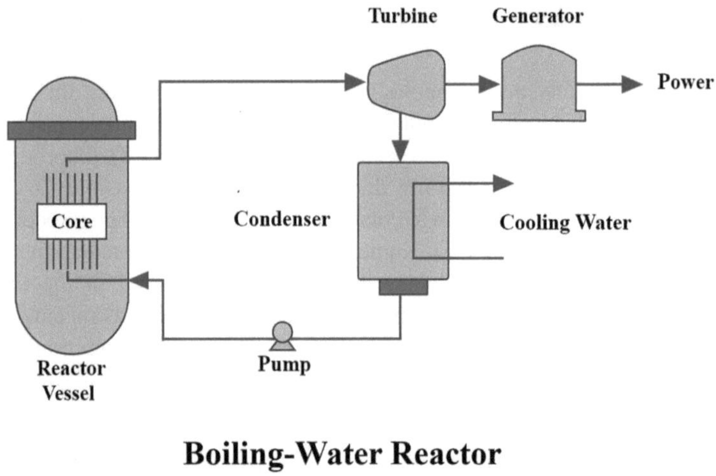

Fig. 11.3 Type of nuclear reactor

energy output is used for water production (Ho et al. 2024). In addition to the aforementioned techniques, it is also possible to integrate seawater desalination systems with nuclear reactors because nuclear desalination has a geographic advantage that could result in significant financial gains. Optimizing the connection between nuclear reactors and desalination units has been the subject of numerous studies (El-Emam et al. 2022; Khamis 2009; Khan and Orfi 2021; Sodaye et al. 2009).

11.5 Nuclear Desalination: The Future of Sustainability and Competitiveness

There are two primary reasons to concentrate on increasing nuclear power's contribution to desalination in the future, both in terms of heat and energy. The first is the above-mentioned growing demand for freshwater, and the second is the growing worry over pollution from fossil fuels and greenhouse gas (GHG) emissions (Lønborg et al. 2024; Buzzetti et al. 2024; Zhang et al. 2024). According to reports, nuclear desalination is generally more cost-effective and competitive than alternatives based on fossil fuels, mostly because of the low fuel cycle costs associated with the former process. According to reports, nuclear desalination is a practical and affordable solution in several parts of the world (El-Emam et al. 2022; Khamis 2009). Since nuclear-assisted desalination plants have advantageous economic and environmental features, they can be regarded as a component of the development of national industrial infrastructure (Ali et al. 2024). For this reason, it is crucial to consider them as strategically significant programs for the welfare of an area or a country rather than just as chemical and thermal process technologies (Sodaye et al. 2009). Nuclear desalination is a significant non-electric use of nuclear heat and power that is closely related to water security, sustainable development, and climate change management (both adaptation and mitigation). Through sector coupling mechanisms, these initiatives will also promote the deployment of a clean hydrogen economy and the sustainable expansion of other water-dependent businesses. They also allow nuclear power plants to operate in a flexible manner in the energy mixes of the future (Misra and Kupitz 2004; El-Ashmawy et al. 2024; Raymond et al. 2024). Different techno-economic and environmental implications result from the energy source used to operate a particular desalination method when the procedures are implemented at scale. In order to facilitate project deployment as part of a long-term water security strategy for countries with nuclear power projects, this section quantifies the most significant of these key characteristics that set nuclear desalination apart from desalination powered by other energy sources (Faibish et al. n.d.; Ozcan et al. 2024; Ho et al. 2024).

11.6 Conclusion

This chapter provided a comprehensive review for the various aspects of nuclear desalination processes. The development of desalination systems and nuclear reactors is rapidly expanding, and many nations are participating. The primary problem in the MENA region is the potential for desalination plant development. Additionally, the usage of desalination systems in nuclear reactors is becoming more accessible as nuclear power generation increases. The application of nuclear reactors in desalination systems is supported by the current situation and a coupled nuclear reactor desalination system. The technological measurement for the production of purified water is strengthened by computational approaches used in the development and integration of desalination systems. In order to create a sustainable and circular water economy, nuclear energy may also be utilized in the future to assist extensive water reuse and recycling initiatives. Therefore, it is necessary to have technology-neutral and supportive regulations that allow the implementation of nuclear power programs and the desalination projects that go along with them.

References

Abdel Monem MAS, Ghandour IAE, Abdel-Sabour MF (2024) Advancing groundwater management and seawater intrusion monitoring in the Middle East and North Africa: exploring the potential of nuclear tracer techniques. In: El-Rawy M, Negm A (eds) Managed aquifer recharge in MENA countries: developments, applications, challenges, strategies, and sustainability. Springer International Publishing, Cham, pp 29–45. https://doi.org/10.1007/978-3-031-58764-1_2

Ali L, Girba I.-A, Pruiu A, Marasescu D (2024) About the maintenance of the heat flow exchangers from energetically naval systems. EBSCOhost. https://openurl.ebsco.com/contentitem/

doi:10.21279%2F1454-864X-16-I1-023?sid=ebsco:plink:crawler&id=ebsco:doi:10.21279%2F1454-864X-16-I1-023. Accessed 11 Nov 2024
Al-Othman A, Darwish NN, Qasim M, Tawalbeh M, Darwish NA, Hilal N (2019) Nuclear desalination: a state-of-the-art review. Desalination 457:39–61. https://doi.org/10.1016/j.desal.2019.01.002
Assia C, Mohamed M (2024) L'eau dans la région de Souss Massa; ressources, gestion et défis. Revue Paysages Géographiques 1(7):7. https://journals.imist.ma/index.php/RPG/article/view/2323. Accessed 9 Nov 2024
Belhaj F, et al (n.d.) Bulletin d'Information Économique de la Région Mena
Benaafi M et al (2024) Combined effects of seawater intrusion and heavy metal pollution on the groundwater resources of Tarout Island, Saudi Arabia. Arab J Sci Eng. https://doi.org/10.1007/s13369-024-09184-2
Bozkaya Ş, Onifade ST, Duran MS (2025) Nuclear energy utilization and the expectations of emission-reduction gains: empirical evidence from economic trajectory of selected utilizing states. Prog Nucl Energy 178:105526. https://doi.org/10.1016/j.pnucene.2024.105526
Buzzetti R, Lo Frano R, Cancemi AS (2024) Feasibility study of desalination plant powered by SMR. Nucl Eng Des 418:112897. https://doi.org/10.1016/j.nucengdes.2023.112897
El-Ashmawy WM, El-Maghlany WM, Elhelw M (2024) Thermo-economic analysis of potential desalination processes utilized by no greenhouse gas emissions power plant. Alex Eng J 109:191–200. https://doi.org/10.1016/j.aej.2024.08.062
El-Emam RS, Ozcan H, Bhattacharyya R, Awerbuch L (2022) Nuclear desalination: a sustainable route to water security. Desalination 542:116082. https://doi.org/10.1016/j.desal.2022.116082
El-Moustaqim K et al (2023a) Combination of microalgae method, decantation, and filtration for domestic wastewater treatment. Sustainability 15(22):22. https://doi.org/10.3390/su152216110
El-Moustaqim K, Mabrouki J, Azrour M (2023b) Monitoring of water toxicity through the Internet of Things to protect the health of the population, pp 85–92. https://doi.org/10.1201/9781003430735-7
El-Moustaqim K, Benchrifa M, Mabrouki J, Hmouni D (2024a) Interaction between the microalgae cultivation system and its effects on nutrient removal from wastewater, pp 70–79. https://doi.org/10.1201/9781003436218-8
El-Moustaqim K, Tallou A, Mabrouki J, Slaoui M, Hmouni D (2024b) Phytoremediation processes for the removal of heavy metals, pp 33–46. https://doi.org/10.1007/978-3-031-55787-3_3
El-Moustaqim K, Mabrouki J, Hmouni D (2025a) Microalgae-based biorefinery: a new approach for wastewater treatment and bioresource collection, pp 315–330. https://doi.org/10.4018/979-8-3693-9450-2.ch018
El-Moustaqim K, Mabrouki J, Hmouni D (2025b) Valorization of agro-industrial wastewater for microalgae cultivation, pp 569–584. https://doi.org/10.4018/979-8-3693-9450-2.ch029
Faibish RS, Konishi T, Gasparini M (n.d.) Application of nuclear energy for seawater desalination: design concepts of nuclear desalination plants. https://doi.org/10.1115/ICONE10-22071
Ho A, Billings BW, Hedengren JD, Powell KM (2024) Flexible operation of nuclear hybrid energy systems for load following and water desalination. Renew Energy Focus 51:100641. https://doi.org/10.1016/j.ref.2024.100641
Kamboj V, Kamboj N, Bisht A (2020) An overview of water quality indices as promising tools for assessing the quality of water resources, pp 188–214. https://doi.org/10.26832/aesa-2020-aepm-013
Khamis I (2009) A global overview on nuclear desalination. Int J Nucl Desalinat 3(4):311–328. https://doi.org/10.1504/IJND.2009.028859
Khan SU-D, Orfi J (2021) Socio-economic and environmental impact of nuclear desalination. Water 13(12):12. https://doi.org/10.3390/w13121637
Lønborg C et al (2024) A global database of dissolved organic matter (DOM) concentration measurements in coastal waters (CoastDOM v1). Earth Syst Sci Data 16(2):1107–1119. https://doi.org/10.5194/essd-16-1107-2024

Misra BM, Kupitz J (2004) The role of nuclear desalination in meeting the potable water needs in water scarce areas in the next decades. Desalination 166:1–9. https://doi.org/10.1016/j.desal.2004.06.053

Ozcan B, Kılıç Depren S, Kartal MT (2024) Impact of nuclear energy and hydro electricity consumption in achieving environmental quality: evidence from load capacity factor by quantile based non-linear approaches. Gondwana Res 129:412–424. https://doi.org/10.1016/j.gr.2023.05.014

Raymond C, Omitaomu OA, Franzese K, Wagner MJ, Lindley B (2024) Siting analysis of a solar-nuclear-desalination integrated energy system. J Nucl Eng 5(3):3. https://doi.org/10.3390/jne5030025

Rich A (2011) Dessalement de l'eau de mer par congélation sur parois froides: aspect thermodynamique et influence des conditions opératoires. PhD thesis, Université Claude Bernard, Lyon I; Université Mohammed V-Agdal (Rabat, Maroc; 1993–2014). https://theses.hal.science/tel-00958034. Accessed 13 Nov 2024

Ritchie H, Rosado P, Roser M (2024) Nuclear energy. Our world in data. https://ourworldindata.org/nuclear-energy. Accessed 7 Dec 2024

Saleh MM, Abdelrazek AA, Mohammed AF, Sokolova EA (2024) A solution to the problem of fresh water shortage in Egypt using nuclear desalination. In: 2024 6th International youth conference on radio electronics, electrical and power engineering (REEPE), pp 1–6. https://doi.org/10.1109/REEPE60449.2024.10479700

Sara L, Nacera L (2015) Dessalement de l'eau de Mer et les Énergies Renouvelables. J Adv Res Sci Technol 2(2):271–279

Sodaye H, Nisan S, Poletiko C, Prabhakar S, Tewari PK (2009) Extraction of uranium from the concentrated brine rejected by integrated nuclear desalination plants. Desalination 235(1):9–32. https://doi.org/10.1016/j.desal.2008.02.005

Tahri T, Bettahar A, Douani M (n.d.) Dessalement de l'eau de mer par l'énergie solaire en vue de l'irrigation d'une serre agricole

Tallou A et al (2024) Phytoremediation of contaminated environments using halophytes: general overview. https://doi.org/10.1007/978-3-031-55787-3_11

Wang P, Tang X, Ren Q (2024) Nuclear power waste heat enhanced salinity-gradient osmotic energy conversion with warm drained seawater. J Power Sources 620:235192. https://doi.org/10.1016/j.jpowsour.2024.235192

Wu Q, Li G, Yin J, Liu M, Yan J, Deguchi Y (2024) The integration of seawater desalination system with nuclear power plant: operational flexibility enhancement and thermo-economic performances. Nucl Eng Des 418:112889. https://doi.org/10.1016/j.nucengdes.2023.112889

Yahyaoui, W, Chibani HE (2024) Etude de la qualité physico-chimique et microbiologique de l'eau potable. Thesis. http://dspace.univ-temouchent.edu.dz/handle/123456789/4941. Accessed 11 Nov 2024

Zhang H et al (2024) Experimental study and model optimization of thermodynamic performance of a single screw water vapor compressor. Int J Refrig 161:135–144. https://doi.org/10.1016/j.ijrefrig.2024.02.024

Zhong Z, Burhan M, Ng KC, Cui X, Chen Q (2024) Low-temperature desalination driven by waste heat of nuclear power plants: a thermo-economic analysis. Desalination 576:117325. https://doi.org/10.1016/j.desal.2024.117325

Chapter 12
A DPSIR Framework to Analyze the Purpose of Using Nuclear Techniques for Marine Environment Protection

Nezha Mejjad and Hamid Amsil

Abstract The health of the world's oceans is under serious threat from various anthropogenic activities, including chemical pollution, ocean acidification, microplastic contamination, eutrophication, and the degradation of blue carbon ecosystems. Multiple analytical techniques, including nuclear techniques, have been developed for monitoring coastal and marine systems. To assess the role of using these techniques for marine ecosystem management, a Driver–Pressure–State–Impact–Response (DPSIR) framework was applied. The analysis underlined the importance of nuclear approaches in providing precise data for marine environmental monitoring. It also emphasized the role of these techniques in bridging scientific capabilities with environmental assessment, policy development, and management strategies.

Keywords DPSIR framework · Nuclear techniques · Ecosystem restoration · Pollution monitoring · Management decisions

12.1 Introduction

Since 1970s, there has been a growing awareness and interest among stakeholders, managers, scientists, and civil society about the conservation of marine and coastal environments, owing to the multiple roles these ecosystems play. This is linked to the fact that, until that time, many communities around the world disposed of their waste directly into the ocean—a practice later prohibited and regulated under the 1972 London Convention and the 1996 London Protocol (Schenker 1973).

N. Mejjad (✉) · H. Amsil
National Centre for Nuclear Energy, Science and Technology (CNESTEN), Rabat, Morocco

N. Mejjad (ed.), *Nuclear-Based Approaches application for marine pollution assessment and monitoring*, Environmental Science and Engineering,
https://doi.org/10.1007/978-3-032-12197-4_12

Since then, the understanding of the importance of these environments has continued to be investigated and studied, as some subfields, such as blue carbon, deep-sea mining, and molecular oceanography, remain emerging fields of research. Indeed, the UNESCO Global Ocean Science Report 2020 underlined that "ocean science is still a relatively poorly funded and underdeveloped field compared to other areas of research and innovation, despite its importance for humanity and human well-being" (IOC-UNESCO 2020). According to the same report (IOC-UNESCO 2020), the share of gross domestic expenditure on research and development (GERD) devoted to ocean science is relatively low, averaging approximately 1.7% of total GERD in 2017, with a range of 0.03% to 11.8%.

Recognizing the environmental issues the ocean faces and its role in sustaining human well-being, there is a need to increase funding for ocean-related projects and programs and to deploy these resources strategically to develop technologies that enable more effective monitoring, conservation, and management.

The health of the world's oceans is under serious threat from multiple stressors, including eutrophication (Vantarakis 2021; Horta et al. 2021), acidification (Miles 2009; Abbasi and Abbasi 2011; Falkenberg et al. 2020), chemical and toxic inputs (Mejjad et al. 2018), microplastics (Mejjad et al. 2023, 2024, 2025), degradation of blue carbon ecosystems (Hossain 2019; Lovelock et al. 2017, 2018), and overfishing (Issifu et al. 2022; Sumaila and Tai 2020). A significant gap remains in understanding the long-term effects of these stressors on the ocean ecosystems and the services they provide (IOC-UNESCO 2020). To bridge this gap, it is crucial to utilize advanced technologies and methods that can produce accurate and precise data to inform management programs, regulations, and policy development.

While several technologies and techniques have been developed and employed to analyze and monitor the stressors threatening the world's oceans, nuclear technologies have emerged as powerful complementary approaches that support existing techniques (Table 12.1) (Zhang et al. 2014). These nuclear techniques offer several advantages, including the ability to reconstruct environmental change over time through radiometric dating of sediment cores, estimate carbon burial rates, and trace the sources of carbon and nutrients.

The integration of these techniques enhances the reliability of environmental assessments and strengthens evidence-based approaches for the sustainable management of coastal and marine environments. This chapter seeks to analyze the purpose of using nuclear techniques for marine environment protection using a Driver (D), Pressures (P), State (S), Impact (I), and Response (R) (DPSIR) framework.

Table 12.1 Main stressors of marine and coastal ecosystems, the purpose of nuclear approaches, and their applications

Environmental issue	Description	Nuclear approaches application
Radioactive pollution	Contamination from natural or anthropogenic sources	**Gamma spectrometry** analysis to measure and quantify radionuclides; **Radiometric dating** to assess historical deposition; **Isotope fingerprinting** for identifying sources
Rare earth elements	Release of REEs from mining, electronics devices, and industrial activities	**Neutron activation analysis (NAA)** for trace detection; **Isotopic techniques** to track deposition and sources
Eutrophication	Nutrient enrichment (N, P) leading to hypoxia, algal blooms, and ecosystem disruption	**Stable isotope analysis** (N-15, C-13) to trace nutrient sources; **Radiotracers** to investigate nutrient cycling
Ocean acidification	Decline in seawater pH due to CO_2 absorption, affecting marine life	**Isotope ratios in carbonates** (C-13, C-14) to analyze carbon cycling and pH history; **Radiocarbon dating** of sediments to establish past acidification $\delta^{11}B$ as paleo-pH proxy
Fish contamination	Bioaccumulation of radionuclides, heavy metals, and organic pollutants in fish	**Gamma spectrometry** for radionuclide measurement; **Isotopic tracing** to define contaminant pathways; **Stable isotopes** for trophic transfer investigations
Blue Carbon ecosystem degradation	Degradation of coastal ecosystems (mangroves, salt marshes, and seagrass) is weakening the capacity for carbon sequestration and storage	**Radiometric dating** (Pb-210 and Cs-137) of sediments to measure carbon burial rates over years; **Stable isotopes (C-13)** to track sources of organic carbon C/N isotopes + radiocarbon for long-term burial
Microplastics pollution	Accumulation of micro- and nano-plastics in water, sediment, and biota	**Radiolabeled tracers** to study transport and fate; **Neutron activation analysis** for associated chemical contaminants
Sediment contamination	Pollution accumulation in river, lake, and coastal sediments	**Radiometric dating (Pb-210, Cs-137)** to reconstruct deposition history; **Isotope fingerprinting** to identify pollutant sources
Climate change impacts	Sea level rise, temperature increase, and ecosystem stress	**Isotope analysis (O-18, H-2)** for paleoclimate reconstruction; **Radiocarbon dating** to track past ecosystem changes ice core isotopes ($\delta^{18}O$, δD) for the global context
Airborne pollution deposition	Transport of pollutants from the atmosphere to the soil and water	**Isotopic tracing** to identify sources; **Radionuclide measurements** to quantify deposition rates
Hydrological changes/ groundwater contamination	Pollution of water resources with heavy metals, radionuclides, or nutrients	**Isotope hydrology (O-18, H-2, Sr isotopes)** to trace water origin, flow, and contaminant transport

(continued)

Table 12.1 (continued)

Environmental issue	Description	Nuclear approaches application
Harmful algal blooms (HABs)	Toxic algae proliferation is affecting marine ecosystems and human health	**Stable isotopes** to track nutrient sources; **Radiotracers** to study nutrient uptake and cycling
Permafrost and carbon release	Thawing permafrost is releasing greenhouse gases	**Radiocarbon dating (C-14)** of organic matter to study carbon release history; **Isotopic analysis** of CO_2 and CH_4 sources

12.2 Methods

12.2.1 DPSIR Framework

The DPSIR framework is widely used in environmental sciences, as it provides a systematic way to link human activities (including economic and social factors), ecological alterations, and management actions. Initially developed by the Organization for Economic Co-operation & Development (OECD 1993), it is now internationally recognized and applied by institutions such as the European Environment Agency (EEA 1999) and the UNEP in 1994 (EEA 1999). Numerous studies have also employed this framework in the marine environment (Huong et al. 2022; Dzoga et al. 2020; Mejjad and Rovere 2021; Mejjad et al. 2022).

In the present study, we used the DPSIR framework to analyze the application of nuclear techniques in marine environment monitoring and management (Fig. 12.1). The framework provides a comprehensive structure, encompassing the entire chain from underlying causes (drivers and pressures) to environmental changes (states and impacts), and ultimately, responses, which include management programs and actions. Importantly, the DPSIR framework helps bridge the gap between science and policy by establishing clear links. For example, sediment dating (state) enables carbon accounting (impact), which in turn supports restoration measures and management strategies (responses). In this way, the framework strengthens the connection between scientific data and policy requirements.

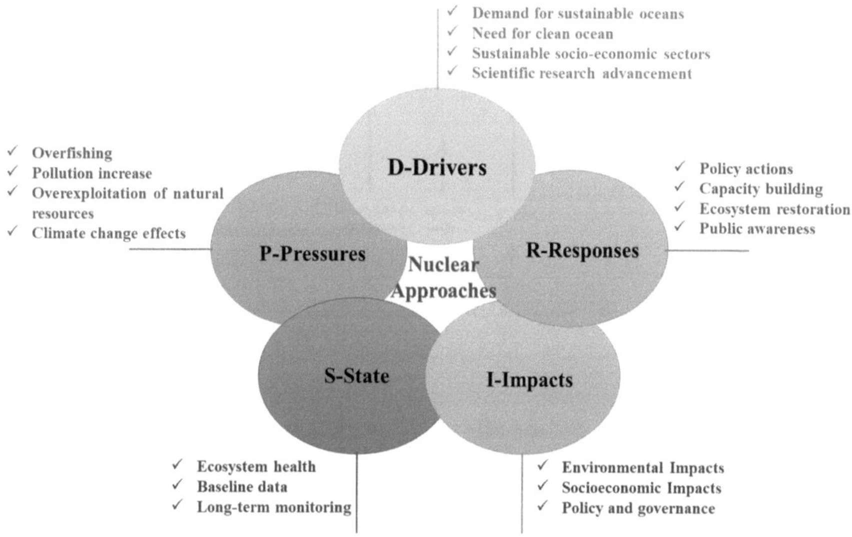

Fig. 12.1 DPSIR Framework applied to nuclear approaches and techniques for marine environment monitoring and management

12.3 Results and Discussions

12.3.1 DPSIR Framework Analysis: Nuclear Techniques in Marine Environment Protection

12.3.1.1 D-Drivers

Drivers for using nuclear techniques in coastal and marine environments include the need for clean and sustainable oceans, the expansion of human activities, and advancements in scientific knowledge (Fig. 12.2). The rising pressures on coastal and marine environments threaten biodiversity, impact socio-economic activities in the sea, and exacerbate the effects of climate change. Consequently, the demand for healthier and more sustainable oceans is a priority, as they provide multiple goods and services, including climate regulation through carbon capture and storage in sediments for millions of years (Arias Ortiz 2019). This process is largely mediated by Blue Carbon Ecosystems (BCEs), which encompass seagrasses, mangroves, and salt marshes. This coastal vegetation helps mitigate the effects of climate change, acting as a nature-based solution (Christianson et al. 2022). However, anthropogenic activities degrade these ecosystems, which may reduce their capacities for carbon storage and threaten stored carbon (Lovelock et al. 2018). This drives scientists to employ nuclear methods to fill knowledge gaps, particularly in estimating carbon burial rates over time, which can only be achieved through such techniques. Similarly, nuclear approaches are applied to study the impacts of toxic chemicals and microplastics, which harm biota and disrupt fisheries and life underwater.

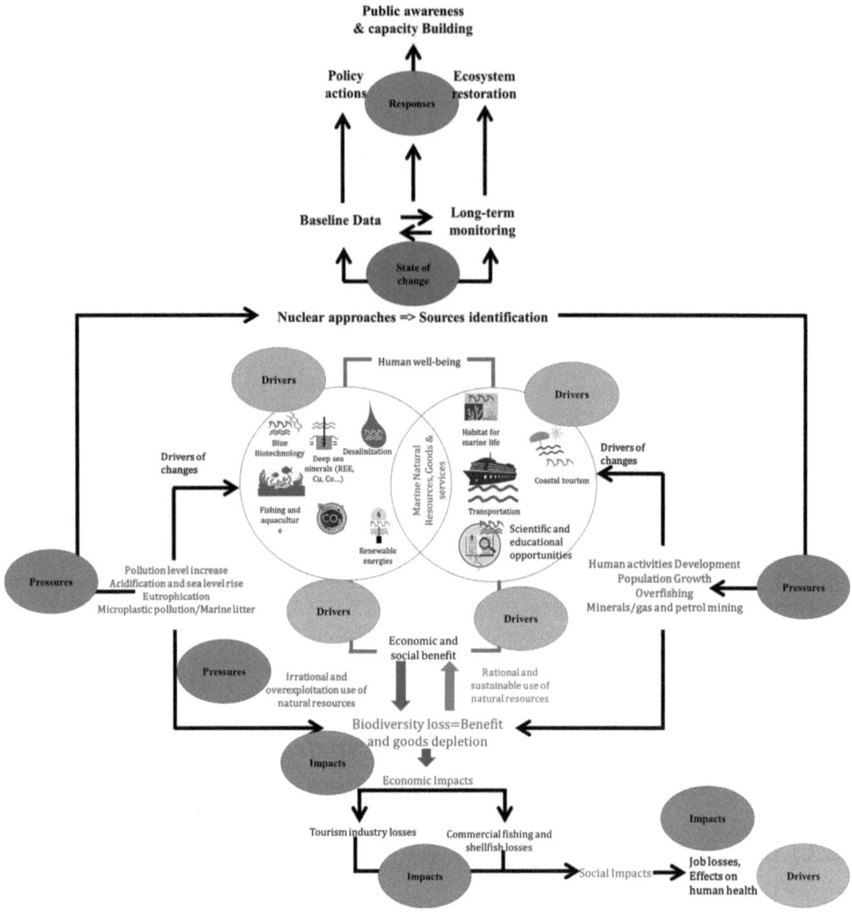

Fig. 12.2 Schematic presentation of primary ocean ecosystem services, Drivers, Pressures, Impacts, and Response related to the application of nuclear approaches

Consequently, they may affect vital economic sectors such as fisheries, aquaculture, and coastal tourism, ultimately leading to job losses (Mejjad 2025). Accordingly, mitigating climate change, ensuring food security, and restoring and conserving coastal and marine environments are the primary drivers for applying nuclear technologies, which are supported by scientific and technological advancements that have made these approaches possible, allowing for precise and long-term assessment and monitoring of environmental change.

12.3.1.2 P-Pressures

Pressures associated with the use of nuclear approaches in coastal and marine environments include pollution and contaminants (such as chemicals, microplastics, radionuclides, and polycyclic aromatic hydrocarbons (PAHs)), land reclamation, and overfishing. The intensification of human activities in coastal and marine environments has led to widespread pollution from chemicals, radioactivity, and microplastics. Industrial processes, chemical manufacturing, agricultural and domestic activities, as well as natural sources, are the main contributors to heavy metal contamination in aquatic systems. Furthermore, the growing demand for mineral resources to support high-technology products has led to an increase in heavy metal releases into coastal and marine areas through riverine input, wastewater, surface runoff, and atmospheric deposition (Hatje et al. 2018).

Human activities are estimated to have amplified global fluxes of toxic heavy metals far beyond natural levels, with lead (Pb) increasing by more than tenfold (Nriagu 1979) and mercury (Hg) by a factor of 3 to 7 (Lamborg et al. 2014), compared with pre-anthropogenic conditions (Hatje et al. 2018). A significant fraction of these toxic metals ultimately accumulates in the ocean (Hatje et al. 2018). Silver (Ag), one of the most poisonous elements for marine biota, has been repeatedly detected in coastal seawater in various regions (Lodeiro et al. 2017), largely due to riverine input and atmospheric deposition from coal combustion, among other sources.

Microplastics, recognized as emerging pollutants, have received increasing attention over recent decades due to their pervasive impacts on biodiversity and nearly all marine ecosystems. Mejjad et al. (2025) reported that microplastics can disrupt and impair the ecological functions of marine and coastal habitats. Approximately 80% of the total waste discarded into the ocean consists of plastics, with more than 13 million metric tons entering coastal and marine environments annually (Pilapitiya and Ratnayake 2024), posing a significant threat to biodiversity and ecosystem services. In addition, microplastics can act as vectors for heavy metals, as they naturally contain elements such as Cd and Zn and can adsorb others, including Cu, Pb, and rare earth elements (REEs) (Liu et al. 2021, 2022; Song et al. 2024).

The increasing release and accumulation of these pollutants in marine ecosystems, combined with the effects of climate change, which include acidification and sea level rise, represent a significant pressure on marine systems. Additionally, overfishing also poses a critical threat, with trawling activities reported to damage habitats such as seagrass beds (Stallings et al. 2014) and coral reefs (Mangi and Roberts 2006; Hall-Spencer et al. 2002). This physical disturbance undermines the role of these ecosystems in climate mitigation by reducing their ability to capture and store carbon. Porz et al. (2024) reported that the use of bottom-contact fishing gear resuspends organic carbon stored in blue carbon sediments, with the fate of this carbon remaining uncertain. The same study's findings revealed high organic carbon loss due to trawling in the specific areas studied, and an increased content of organic carbon was recorded in untrawled zones (Porz et al. 2024). However, Zhang et al. (2024) observed that both loss and gain of sedimentary organic carbon may occur in lightly trawled sites, but heavily trawled areas consistently showed net depletion.

Such pressures highlight the urgent need for methodological and technical tools, such as nuclear techniques, that can bridge the gap between scientific evidence, policy, and management. Applying nuclear approaches supports environmental preservation, climate change mitigation, and helps reduce the socio-economic impacts of declining fish stocks and pollution, which threaten activities dependent on ocean health.

12.3.1.3 S-State

The characterization of environmental change is made possible through the use of nuclear techniques. Radiometric dating methods (Pb-210, Cs-137, and C-14) allow the assessment of ecosystem states by tracking changes over time and identifying pollutant sources through historical deposition records. These methods are also used to estimate sediment accumulation and carbon burial rates. Such datasets are crucial for linking changes in pollutant levels with the development of human activities and natural events in the study area. In turn, this information supports the design of mitigation and restoration projects, as well as the development of effective regulations and policies.

Stable isotopes ($\delta^{13}C$, $\delta^{15}N$) further contribute to assessing ecosystem conditions by tracing nutrient pathways, identifying pollution sources, and analyzing food-web dynamics. Together, these approaches establish baseline datasets and reconstruct the history of environmental change, providing a robust foundation for understanding the long-term dynamics and resilience of coastal and marine ecosystems.

12.3.1.4 I-Impacts

The impact of using nuclear techniques is multifaceted and operates across various levels (Fig. 12.2).

Scientific level: They enable the reconstruction of the history of deposition of heavy metals, radionuclides, microplastics, and other pollutants. This allows establishing age–depth relationships, which help reconstruct pollution history and contaminant inputs over time, permit identifying periods of increased contamination levels, track variations in pollutant sources, and assess ecosystem vulnerability. This is of great interest for understanding past and present variations of pollutants, quantifying enrichment degrees, and identifying potential causes and sources, which supports coastal and marine environment restoration programs and management decisions. Additionally, these approaches are widely used to evaluate the past and present capacities of blue carbon ecosystems for carbon storage, through the estimation of carbon accumulation rates, as well as by assessing variations in carbon stocks over depth and time.

Socio-economic level: Data generated through these approaches, and mainly radiometric dating, support the advancement of blue carbon markets and inform key stakeholders in various sectors that rely on ocean resources, including fisheries, coastal tourism, and coastal management. Such data also contribute to enhancing food safety monitoring, as radioanalytical methods are used in seafood safety to detect the transfer of radioactive contaminants from the environment to the marine food chain in the event of a nuclear accident. Moreover, measuring radioactive element concentrations in seafood helps assess radiation exposure risks for consumers, thereby protecting human health by preventing unsafe products that exceed permissible limits and standards from reaching the market. This, in turn, impacts socio-economic sectors, mainly fisheries, coastal tourism, and all related economic activities that rely on seafood and coastal marine resources.

Governance level: Nuclear techniques provide precise data and evidence-based inputs for international programs and policy frameworks such as the UN Sustainable Development Goals (SDGs) (Leite de Almeida et al. 2021) and the Paris Agreement. They contribute to risk assessment, regulatory development, and environmental protection measures. For example, the IAEA has established safety guidance for ecological protection from ionizing radiation (e.g. International Atomic Energy Agency 2010, 2023). In this way, nuclear approaches directly inform governance, shaping both national and international strategies for environmental management and sustainability.

12.3.1.5 R-Responses

Based on the drivers, pressures, and impacts of applying nuclear approaches, numerous responses can be developed (Table 12.2). Evidence derived from radioanalytical approaches supports pollution regulation, marine spatial planning, and ecosystem restoration programs. For example, the IAEA informs the regulatory framework established under the London Convention and strengthens international rules against the dumping of radioactive waste at sea. IAEA experts apply radioanalytical techniques and risk assessment approaches in conjunction with the "de minimis" concept, which allows only materials with an insignificant radiological impact to be considered for disposal. Additionally, by evaluating the maximum amount of radiation exposure that key reference species (e.g., flatfish, brown seaweed, and crab) can receive without showing biological effects, threshold values have been established (International Atomic Energy Agency 2015). This demonstrates how the application of nuclear approaches provides a scientific foundation for regulation, environmental assessment, and policy development.

At the 2025 United Nations Ocean Conference in Nice, scientists from the IAEA highlighted the role of nuclear technology in addressing three interconnected planetary crises: pollution, climate change, and biodiversity loss, thereby supporting the governance and regulation of marine ecosystems (IAEA 2025). The NUTEC Plastics initiative presents an exemplary model of actions through which the IAEA provides countries with methodologies, protocols, and tools to monitor and quantify microplastics in the ocean, enabling the production of comparable data that informs regulatory frameworks, policies, and international negotiations on plastic pollution in the sea. This initiative combines nuclear radioanalytical measurement and radiation application to assess and quantify microplastics over time, and to establish recycling solutions. Over 100 laboratories in 99 countries participating in the IAEA's NUTEC Plastics initiative are equipped with the equipment and tools

Table 12.2 DPSIR framework of environmental issues in marine environments and the role of nuclear techniques

D—Drivers	P—Pressures	S—State	I—Impacts	R—Responses
Need for healthy and sustainable oceans	Pollution (radionuclides, heavy metals, PAH, microplastics)	Degraded ecosystems	Biodiversity loss	Pollution regulation and governance
Growth of anthropogenic activities in coastal areas	Coastal pollution, land reclamation, and tourism infrastructure	Altered water chemistry	Biodiversity decline, erosion, and greater vulnerability of the coast to storms	Marine spatial planning
Climate change mitigation (blue carbon ecosystems)	Mangrove, seagrass losses	Loss of carbon storage capacity	Higher CO_2 emissions, reduced climate resilience, and socio-economic losses for coastal communities	Ecosystem restoration programs
Biodiversity and ecosystem protection	Eutrophication and acidification	Carbon loss, ecosystem imbalance	Coastal vulnerability and ecosystem degradation	International conventions and rules (e.g., London Convention)
Food security and socio-economic reliance (tourism, fisheries, aquaculture)	Contaminants in seafood (algal toxins, fish contamination)	Decreased water quality	Risks to communities' health and economic costs	Risk assessment and thresholds for species
Scientific and technological advancements enabling isotopic/nuclear applications	Combined planetary crises (climate change, pollution, biodiversity loss)	Long-term variation is measurable through nuclear techniques	Global environmental stress	International programs (IAEA OA-ICC, NUTEC Plastics)

essential for plastic quantification. Meanwhile, 52 countries worldwide are working on projects aimed at developing innovative recycling technology that contributes to mitigation measures.

Besides, numerous training programs have been established to build capacities in the field of oceanography and the application of nuclear techniques, as well as to develop networks and promote collaborations among scientific institutes and universities worldwide. Among the established programs, there are the IAEA Ocean Acidification International Coordination Centre (OA-ICC) (Dupont et al. 2025) and the NUTEC Plastics Program. Both programs address major environmental planetary issues, including marine pollution and acidification. These initiatives include developing shared methodological protocols for sampling and analysis to ensure the provision of comparable data. Additionally, they aim to enhance capacities, expertise, and practical knowledge by training scientists from various scientific institutes and universities worldwide. This enables the improvement of national monitoring programs for ocean acidification and marine pollution, promoting international collaboration to enhance understanding of both ocean health and establish resilience in vulnerable coastal communities, and ensuring the sustainable development and use of marine resources.

The IAEA OA-ICC program was launched in response to growing concerns about ocean acidification among IAEA member states, as expressed during the UN Rio+20 conference in 2012 (Dupont et al. 2025). As key results, this program hosted 31 training related to ocean acidification (OA) that known the participation of more than 500 participants from 108 countries (Dupont et al. 2025), among which some have become recognized experts in their home countries as well as at regional and international levels and are contributing to the scientific communities by publishing high quality papers on OA (e.g., Edworthy et al. 2022).

Additionally, the application of nuclear approaches to address environmental issues in the coastal and marine environment contributes directly and indirectly to achieving the 17 Sustainable Development Goals. Providing data related to coastal water quality, tracking the origin of pollution, and quantifying pollutants enables evidence-based inputs that can contribute to achieving the stated goals.

12.4 Conclusions

There are various ocean stressors, including microplastic pollution, chemical and biological contamination, eutrophication, deoxygenation, acidification, degradation of blue carbon ecosystems, and overfishing. Sustained pressures can severely damage coastal and marine environments, negatively affecting the livelihoods and well-being of coastal communities.

This analysis highlighted the role of nuclear methods in accurately monitoring pollution, minimizing associated risks, and mitigating their potential effects on biodiversity and coastal communities. Additionally, the deployment of nuclear approaches for assessing pollution and analyzing various stressors, as well as their potential interactions, has helped inform management decisions and policy development.

At the economic level, these approaches have contributed to avoiding or minimizing job losses caused by coastal resource depletion or degradation, ensuring food safety by providing data on seafood quality, while also helping establish standards and permissible limits for their consumption.

At the social level, the application of nuclear approaches has been accompanied by capacity-building initiatives. For example, the IAEA has supported multiple projects, programs, and actions related to the application of nuclear-based techniques. This intergovernmental organization, since its founding in 1957, has supported regional and international cooperation, coordination, and harmonization in the monitoring field by generating safety standards and recommendations for reactor safety, nuclear waste management, radiological protection, emergency preparedness, the transport of radioactive substances, and the security of nuclear materials and facilities. This was achieved through providing training and expert guidance to countries worldwide to help them use nuclear techniques. Additionally, by supporting research, innovation, and development in the field, numerous research projects were conducted in countries worldwide, where nuclear approaches were applied to address public health and environmental issues.

Accordingly, the DPSIR analysis has defined the role that nuclear approaches play in advancing knowledge of marine environmental issues, bridging gaps, and delivering datasets crucial for implementing policy decisions, restoration, and management strategies, which evidently have positive social, economic, environmental, and regulatory impacts.

References

Abbasi T, Abbasi SA (2011) Ocean acidification: the newest threat to the global environment. Crit Rev Environ Sci Technol 41(18):1601–1663

Arias Ortiz A (2019) Carbon sequestration rates in coastal blue carbon ecosystems: a perspective on climate change mitigation

Christianson AB, Cabré A, Bernal B, Baez SK, Leung S, Poloczanska E (2022) The promise of blue carbon climate solutions: where the science supports ocean-climate policy. Front Mar Sci 9:851448

Dupont S, Edworthy C, Sánchez-Noguera C, Metian M, Friedrich J, Flickinger S, Bantelman A, Galdino C, Graba F, Anghelici O, Hansson L (2025) The IAEA Ocean Acidification International Coordination Centre Capacity Building Program: empowering member states to address and minimize the impacts of ocean acidification. Oceanography 38(1):111–113

Dzoga M, Simatele DM, Munga C, Yonge S (2020) Application of the DPSIR framework to coastal and marine fisheries management in Kenya. Ocean Sci J 55(2):193–201

Edworthy C, Potts WM, Dupont S, Duncan MI, Bornman TG, James NC (2022) A baseline assessment of coastal pH variability in a temperate South African embayment: implications for biological ocean acidification research. Afr J Mar Sci 44:367–381

EEA (1999) Towards environmental pressure indicators for the EU, 1st edn. European Environment Agency

Falkenberg LJ, Bellerby RG, Connell SD, Fleming LE, Maycock B, Russell BD et al (2020) Ocean acidification and human health. Int J Environ Res Public Health 17(12):4563

Hall-Spencer J, Allain V, Fosså JH (2002) Trawling damage to Northeast Atlantic ancient coral reefs. Proc R Soc Lond B Biol Sci 269(1490):507–511

Hatje V, Lamborg CH, Boyle EA (2018) Trace-metal contaminants: human footprint on the ocean. Elements 14:403–408

Horta PA, Rörig LR, Costa GB, Baruffi JB, Bastos E, Rocha LS et al (2021) Marine eutrophication: overview from now to the future. In: Anthropogenic pollution of aquatic ecosystems, pp 157–180

Hossain MM (2019) Future importance of healthy oceans: ecosystem functions and biodiversity, marine pollution, carbon sequestration, ecosystem goods and services. J Ocean Coast Econ 6(2):4

Huong DTT, Ha NTT, Do Khanh G, Van Thanh N, Hens L (2022) Sustainability assessment of coastal ecosystems: DPSIR analysis for beaches at the Northeast Coast of Vietnam. Environ Dev Sustain 24(4):5032–5051

IAEA (2025) Nuclear techniques make waves at UN ocean conference. IAEA

International Atomic Energy Agency (2010) Environmental and source monitoring for purposes of radiation protection. Safety guide (Spanish edn). IAEA, Vienna
International Atomic Energy Agency (2015) Inventory of radioactive material resulting from historical dumping, accidents and losses at sea. IAEA, Vienna
International Atomic Energy Agency (2023) Radioprotection du public et de l'environnement. IAEA, Vienna
IOC-UNESCO (2020) Global ocean science report 2020—charting capacity for ocean sustainability. UNESCO, Paris
Issifu I, Alava JJ, Lam VW, Sumaila UR (2022) Impact of ocean warming, overfishing and mercury on European fisheries: a risk assessment and policy solution framework. Front Mar Sci 8:770805
Lamborg CH, Hammerschmidt CR, Bowman KL, Swarr GJ, Munson KM, Ohnemus DC et al (2014) A global ocean inventory of anthropogenic mercury based on water column measurements. Nature 512(7512):65–68
Leite de Almeida CM, Bergqvist E, Thacker S et al (2021) Actions to align energy projects with the Sustainable Development Goals. Discov Sustain 2:16
Liu S, Shi J, Wang J, Dai Y, Li H, Li J, Liu X, Chen X, Wang Z, Zhang P (2021) Interactions between microplastics and heavy metals in aquatic environments: a review. Front Microbiol 12:652520
Liu S, Huang J, Zhang W, Shi L, Yi K, Yu H, Zhang C, Li S, Li J (2022) Microplastics as a vehicle of heavy metals in aquatic environments: a review of adsorption factors, mechanisms, and biological effects. J Environ Manag 302:113995
Lodeiro P, Browning TJ, Achterberg EP, Guillou A, El-Shahawi MS (2017) Mechanisms of silver nanoparticle toxicity to the coastal marine diatom Chaetoceros curvisetus. Sci Rep 7(1):10777
Lovelock CE, Atwood T, Baldock J, Duarte CM, Hickey S, Lavery PS et al (2017) Assessing the risk of carbon dioxide emissions from blue carbon ecosystems. Front Ecol Environ 15(5):257–265
Lovelock CE, Friess DA, Kauffman JB, Fourqurean JW (2018) Human impacts on blue carbon ecosystems. In: A blue carbon primer. CRC Press, pp 17–24
Mangi SC, Roberts CM (2006) Quantifying the environmental impacts of artisanal fishing gear on Kenya's coral reef ecosystems. Mar Pollut Bull 52(12):1646–1660
Mejjad N (2025) Marine environmental quality degradation: socioeconomic and environmental implications. In: Leal Filho W, Salvia AL, Eustachio JPP, Dinis MAP (eds) Handbook of sustainable blue economy. Springer, Cham
Mejjad N, Rovere M (2021) Understanding the impacts of blue economy growth on deep-sea ecosystem services. Sustainability 13(22):12478
Mejjad N, Laissaoui A, El-Hammoumi O, Fekri A, Amsil H, El-Yahyaoui A, Benkdad A (2018) Geochemical, radiometric, and environmental approaches for the assessment of the intensity and chronology of metal contamination in the sediment cores from Oualidia lagoon (Morocco). Environ Sci Pollut Res 25(23):22872–22888
Mejjad N, Rossi A, Pavel AB (2022) The coastal tourism industry in the Mediterranean: a critical review of the socio-economic and environmental pressures & impacts. Tour Manag Perspect 44:101007
Mejjad N, Laissaoui A, Fekri A, Hammoumi OE (2023) Marine plastic pollution in Morocco: state of the knowledge on origin, occurrence, fate, and management. Environ Sci Pollut Res 30(49):107371–107389
Mejjad N, Bououarour O, Laissaoui A, Loukili I, Benmansour M (2024) Fish as a bioindicator of microplastic contamination in the Mediterranean Sea: state of knowledge. Royal Society of Chemistry
Mejjad N, Safhi AEM, Laissaoui A (2025) Insightful analytical review of potential impacts of microplastic pollution on coastal and marine ecosystem services. J Hazard Mater Adv 17:100578
Miles EL (2009) On the increasing vulnerability of the world ocean to multiple stresses. Annu Rev Environ Resour 34:17–41

Nriagu JO (1979) Global inventory of natural and anthropogenic emissions of trace-metals to the atmosphere. Nature 279:409–411

OECD (1993) OECD Core set of indicators for environmental performance reviews: a synthesis report by the Group on the State of the Environment. OECD Environ Monogr 83:OECD/GD(93)179

Pilapitiya PNT, Ratnayake AS (2024) The world of plastic waste: a review. Cleaner Mater 11:100220

Porz L, Zhang W, Christiansen N, Kossack J, Daewel U, Schrum C (2024) Quantification and mitigation of bottom-trawling impacts on sedimentary organic carbon stocks in the North Sea. Biogeosciences 21(10):2547–2570

Schenker MS (1973) Saving a dying sea—the London Convention on ocean dumping. Cornell Intl LJ 7:32

Song D, Shang Y, Wang X et al (2024) Microplastic and heavy metals distributions in urban Rivers sediments, China. J Ocean Univ China 23:1015–1025

Stallings CD, Brower JP, Heinlein Loch JM, Mickle A (2014) Commercial trawling in seagrass beds: bycatch and long-term trends in effort of a major shrimp fishery. Mar Ecol Prog Ser 513:143–153

Sumaila UR, Tai TC (2020) End overfishing and increase the resilience of the ocean to climate change. Front Mar Sci 7:523

Vantarakis A (2021) Eutrophication and public health. In: Chemical lake restoration: technologies, innovations and economic perspectives. Springer International Publishing, Cham, pp 23–47

Zhang ZY, He X, Ma YH, Ding YY, Chai ZF (2014) New nuclear facilities and their analytical applications in China. Radiochim Acta 102(1–2):3–11

Zhang W, Porz L, Yilmaz R, Wallmann K, Spiegel T, Neumann A, Holtappels M et al (2024) Long-term carbon storage in shelf sea sediments reduced by intensive bottom trawling. Nat Geosci 17(12):1268–1276

MIX
Papier aus verantwortungsvollen Quellen
Paper from responsible sources
FSC® C105338

If you have any concerns about our products,
you can contact us on
ProductSafety@springernature.com

In case Publisher is established outside the EU,
the EU authorized representative is:
Springer Nature Customer Service Center GmbH
Europaplatz 3, 69115 Heidelberg, Germany

Printed by Libri Plureos GmbH
in Hamburg, Germany